Contents

I. Introduction 1
  1. The Role of Radiation Workers 1

II. Scientific Basis 3
  1. Structure of Atom 3
  2. Structure of Atomic Nucleus 5
  3. Isotopes 6
  4. Radioisotopes and Radioactivity 8
  5. Half-Life of a Radioisotopes 9
  6. Types of Radioactive Decay 11
  7. X-Rays 19
  8. Accelerators 21
  9. Types of Radiation 22
  10. Quantities and Units Related to Radiation 24

III. Safe Handling 29
  1. Who is the Radiation Protection Staff ? 29
  2. Protection Against External Exposure 31
  3. Protection Against Internal Exposure 33
  4. Attitude when Handling Radioisotopes and Radiation 37
  5. Radiation Monitoring 39
  6. Education and Training 47
  7. From Procurement to Waste management of Radioisotopes 49
  8. Usage of Radioisotopes and Radiation Generators 51
  9. Storage of Radioisotopes 53
  10. Discharge of radioisotopes from facilities and radioactive waste management 55
  11. Record Keeping(Radioisotopes tracking) 57
  12. Record Keeping(Dose) 59

13. Medical Examinations 61
14. Procedures at the Time of an Accident or Emergency 63
15. Procedures in the Event of Excess Exposure or Contamination 65

IV. Biological Effects of Radiation 67
  1. Classification of Radiation Effects
      on the Human Body 67
  2. Acute Effects 68
  3. Late Effects 69
  4. Heritable Effects 70
  5. Biological Effects Depend on the Dose Received 71
  6. Effects from External and Internal Exposures 73
  7. What is Equivalent Dose? 75
  8. What is Effective Dose? 77
  9. Assessment of Internal Dose 80
  10. Natural Radiation and Artificial Radiation 82
  11. Radiation and Health Effects 84
  12. Exposure Categories 86
  13. Risk 88
  14. Limits on Personal Exposure 90

V. The Act and relevant Ordinances 93
  1. Act on Prevention of Radiation Hazards due to Radioisotopes, etc. 93
  2. Hierarchy of the Act and relevant Ordinances 95
  3. Laws and Ordinances Concerning Radiation 97
  4. What is the Act on Prevention
      of Radiation Hazards due to Radioisotopes, etc.? 99
  5. Laws and Ordinances to Protect Workers 101
  6. Who are Radiation Workers? 103
  7. Who are Radiation Protection Supervisors? 105
  8. What is Radiation under Laws and Ordinances? 107
  9. What are Radioisotopes? 109
  10. What are Radiation Generators? 110

11. What are Sealed Radioisotopes? 111
12. What is the Controlled Area? 113
13. What is meant by
    "Radiation Levels outside Controlled Areas"? 115

VI. Radiation Hazards Prevention Program 117
  1. What is the Radiation Hazards Prevention Program? 117

> **Note**
> This text is for beginners who are to work in laboratories using radiation and radioisotopes. Chapters I to IV contain scientific and technical matters for radiation protection based on the internationally common understandings, while Chapters V and VI are compliant with Japanese Laws and Ordinances concerning radiation protection, revised in January, 2016.

Ⅰ. Introduction

1. The Role of Radiation Workers

The purpose of "the Act on Prevention of Radiation Hazards due to Radioisotopes, etc.", with relevant ordinances, cabinet order and notifications etc. (hereinafter referred to as "the Act and relevant Ordinances"), is not only to minimize the radiation exposure of individual workers and the public, but also to ensure the safety of people by preventing contaminations in both the workplace and the general environment. To satisfy this purpose, it goes without saying that facilities and equipments should be well maintained. But there are many areas where little can be achieved without the cooperation of the radiation workers themselves. That is to say, it is difficult to attain the purpose of the Law without appropriate handling of radiation and radioisotopes by radiation workers, no matter how much effort is put into maintaining facilities and equipments. Being aware of how to protect themselves, as well as the significance of their role in radiation protection management, each radiation worker must cooperate with others at his or her workplace.

1-1. What knowledge must radiation workers have?

In order to deal with radiation and radioactive materials safely at their workplace and to attain the goal of radiation protection, radiation workers must understand at least the following:

1) Basic provisions of the Act on Prevention of Radiation Hazards due to Radioisotopes, etc.;
2) Basic physics of the radioactive materials that they handle;
3) Biological effects of radiation;
4) Techniques for preventing contamination;
5) Methods for radioactive waste management; and
6) How to deal with abnormal and emergency situations.

I. Introduction

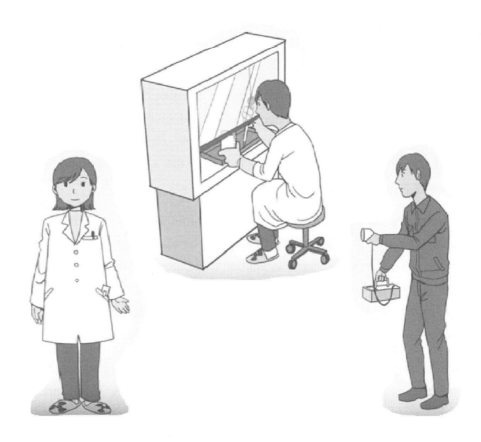

Fig. 1 What knowledge must radiation workers have?

II. Scientific Basis

1. Structure of Atom

The structure of an atom is similar to that of the solar system.

All objects are composed of **atoms**. At the center of each atom is a nucleus with a positive electric charge, around which electrons with the negative charge are moving along fixed orbitals. These electrons are called **orbital electrons**. This structure can be compared to the solar system, where planets are orbiting around the sun.

Normally, the positive charge of the atomic nucleus is equal to the total negative charge of the orbital electrons, and the atom as a whole is electrically neutral. When external energy is applied to an atom, an orbital electron may be kicked out of its orbital, becoming a free electron no longer bound to the nucleus. The atom then remains as a positively charged **ion**. This phenomenon is called **ionization**.

Sometimes, the application of external energy may cause an orbital electron to move to an orbital further from the nucleus, leaving the nearer orbital position vacant. Such an atom is said to be **excited**.

The diameter of an atomic nucleus ranges from 10^{-15} to 10^{-14} m, with the orbitals of the electrons extending much further out. The size of the entire atom is of the order of 10^{-10} m. If, for example, the atomic nucleus is pictured as being the size of a tennis ball, then the orbital electrons would be as far as 5 kilometers away.

II. Scientific Basis

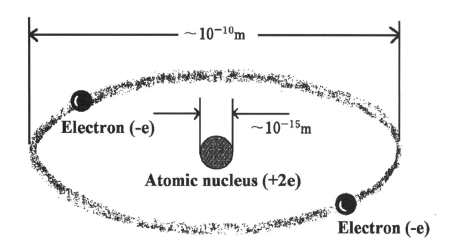

Fig. 2 Structure of a helium (He) atom: The charge (e) on each electron is called elementary electric charge. The helium nucleus has a positive charge twice that of the elementary electric charge, balancing the two orbital electrons.

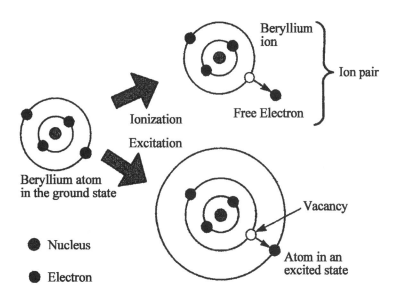

Fig. 3 Ionization and excitation of a beryllium atom

2. Structure of Atomic Nucleus

The atomic nucleus consists of two kinds of particles, protons and neutrons.

The atomic nucleus consists of protons, which are positively charged, and neutrons, which have no charge. The positive charge of a proton is equal, but opposite, to the negative charge of an electron. The mass of a proton is approximately 1,840 times that of an electron; the mass of a neutron is almost the same as, but slightly larger than, that of a proton. Thus, because an electron is so much lighter than either a proton or a neutron, the mass of an atom can be deemed to be essentially equal to the mass of its nucleus.

The particles that make up the nucleus, i.e. the protons and neutrons, are called nucleons. Nucleons are held together by what is called the nuclear force, which overcomes the naturally repulsive tendency of the positively charged protons.

The number of protons in a nucleus is normally the same as the number of orbital electrons around the nucleus. The chemical nature of an atom is determined by the number of protons in its nucleus. This number is called the **atomic number**, and it defines a chemical element.

Atomic number = number of protons in an atomic nucleus
= number of orbital electrons in a neutral atom

Fig. 4 Just as a drop of water looks like a single structure but actually consists of a great number of molecules, the atomic nucleus is composed of individual protons and neutrons.

II. Scientific Basis

3. Isotopes

Isotopes are atomic brothers.

The type of an atomic nucleus is determined by the number of its protons and neutrons. When two atoms have the same number of protons but a different number of neutrons, they have different masses, but the same chemical nature. That is to say, they are the same element.   These atoms are like brothers, and are called **isotopes**.

> In order to distinguish among isotopes, the total number of protons and neutrons (called the **mass number**) is shown at the upper left of the element's symbol. The atomic mass is roughly expressed by the mass number.

Mass number = number of protons + number of neutrons

Hydrogen (H, atomic number 1) is an element with one proton, and exists in three isotopes: ^1H, ^2H and ^3H. ^2H is called deuterium (D), also known as heavy hydrogen, and ^3H is called tritium (T).

In elements having smaller atomic number, light elements, the number of neutrons is more or less equal to the number of protons, but, in elements having larger atomic number, heavy elements, the number of neutrons exceeds the number of protons. For example, the nucleus of ^{16}O consists of eight protons and eight neutrons, but the nucleus of ^{226}Ra (atomic number 88) has 88 protons and 138 neutrons.

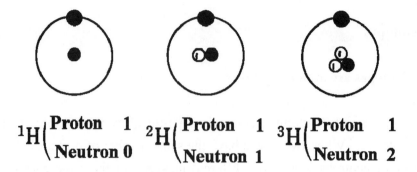

Fig. 5 Hydrogen isotopes: In all cases, the nucleus has one proton.

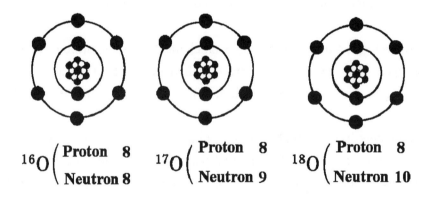

Fig. 6 Oxygen isotopes: In all cases, the nucleus has eight protons.

II. Scientific Basis

4. Radioisotopes and Radioactivity

Isotopes that spontaneously give off radiation and whose nuclei decay into other types of nuclei are called radioisotopes. Instead of the "radioisotope", the terms "radionuclide" or "radioactive nuclide" are often used.

Among isotopes, there are some in which the nucleus emits radiation and is converted into another type of nucleus spontaneously, i.e. without the application of any external condition, such as pressure, temperature, chemical treatment, etc. These are called **radioisotopes**. This property is called **radioactivity**, and the conversion of the nucleus is called radioactive **decay** or **disintegration**. Emitted radiation can be in the form of alpha rays, beta rays, gamma rays, or others. The nucleus before decay is called the **parent**, and the one after decay is called the **daughter**.

Approximately 70 kinds of radioisotopes exist in nature. These include uranium, thorium, radium, and potassium (^{40}K). In addition, there are more than 2,000 kinds of radioisotopes that have been created artificially using nuclear reactors or accelerators.

The term "**activity**" is used to describe the magnitude of radioactivity, the unit for which is the **becquerel (Bq)**, the number of decay events per second.

$$1 \text{ becquerel} = 1 \text{ decay event per second}$$

The old unit of activity **curie (Ci),** which was defined originally based on the activity of 1 gram of ^{226}Ra, is related to the Becquerel as shown in the following equations:

$$1 \text{ Ci} = 3.7 \times 10^{10} \text{ Bq}$$

$$1 \text{ Bq} = 0.2702... \times 10^{-10} \text{ Ci} \approx 27 \text{ pCi}$$

5. Half-Life of a Radioisotopes

A radioisotope has an inherent half-life.

Let us think of a group of atoms of a radioisotope. As the individual atoms decay (i.e., as their nuclei decay), they are converted into another type of atom. The time required for the number of atoms of the radioisotope to become half their original number, namely, for half of them to be converted into another type of atom, is called the "**half-life**". The activity of a certain amount of the radioisotope is in proportion to the number of original atoms remaining.

Half-life is inherent to a radioisotope, and is not affected by external conditions, such as temperature or pressure. Half-lives vary among the kinds of radioisotopes, from more than several billion years to only a very small fraction of a second.

Fig. 7 shows the decrease in activity over time. Activity measured as 1 MBq becomes 0.5 times of 1 MBq after one half-life; 0.25 times of 1 MBq after two half-lives; and 0.125 times of 1 MBq after three half-lives.

II. Scientific Basis

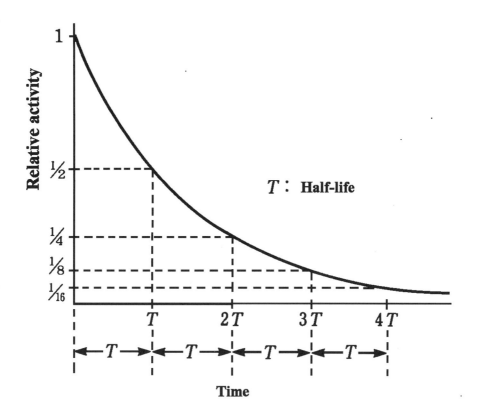

Fig. 7 Decrease in activity: Activity is reduced by half with the passage of each half-life (*T*).

6. Types of Radioactive Decay

6-1. Alpha rays are fast helium nuclei emitted from atomic nuclei.

In one type of radioactive decay, a set of two protons and two neutrons, i.e. a helium nucleus, is emitted at high speed from a heavy nucleus consisting of a large number of protons and neutrons, such as uranium or thorium. Such emissions are called alpha (α) rays, which are said to consist of alpha particles. After an alpha particle is emitted, there remains an atomic nucleus with two protons and two neutrons less than the original nucleus (see Fig. 8). This type of decay is called **alpha decay**.

6-2. Beta rays are fast electrons emitted from atomic nuclei.

In another type of radioactive decay, a neutron changes to a proton in the nucleus, and an electron is emitted. Such emissions are called beta (β) rays, which consist of beta particles. After a beta particle is emitted, there remains a nucleus with one proton more and one neutron less than the original nucleus (see Fig. 9). This type of decay is called **beta decay**.

The words "decay" and "disintegration" suggest that the atomic nucleus is destroyed or falls into pieces. Actually, the nucleus is merely converted into a different kind of nucleus.

II. Scientific Basis

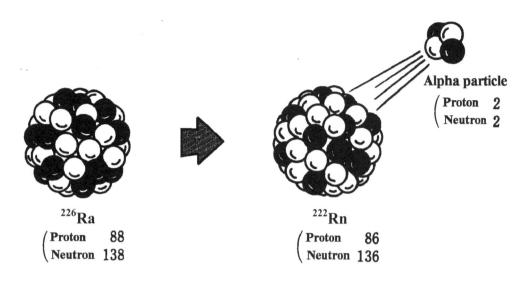

Fig. 8
Decay of ^{226}Ra(radium-226): ^{226}Ra(radium-226) becomes ^{222}Rn(radon-222) by emitting an alpha particle.

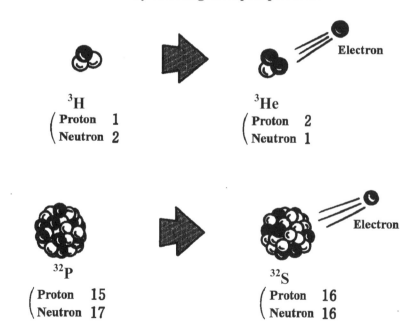

Fig. 9 Decay of ^{3}H (tritium) and ^{32}P (phosphorus-32): ^{3}H changes to ^{3}He (helium-3) by emitting beta particles, and ^{32}P changes to ^{32}S (sulphur-32) by emitting beta particles. In beta decay, one neutron in the nucleus changes into a proton and an electron, and the electron is emitted.

6-3. A positron sometimes emerges from a nucleus.

In another type of radioactive decay, a proton in the nucleus changes into a neutron and a positron (or a positive electron) is emitted. Positrons are described as beta-plus (β^+) rays, and this type of decay is called positron decay or β^+ decay* (see Fig. 10). Positrons have the same mass as electrons, and the same but opposite electric charge. After a positron is emitted, there remains a nucleus with one proton less and one neutron more than the original nucleus. Positrons are attracted easily to any available negative electron, whereupon the electron and the positron—being in fact the anti-particles of each other—annihilate. This annihilation process produces electromagnetic waves (two photons of 0.51 MeV each) emitted in opposite directions with energy equal to the mass of the electron pair (see Fig. 11). Such waves are called annihilation radiation.

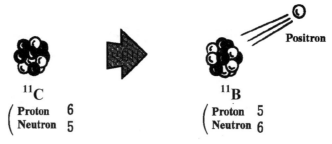

Fig. 10 Positron decay of ^{11}C (carbon-11) : ^{11}C becomes ^{11}B (boron-11) by emitting a positron. In positron decay, one proton in the atomic nucleus changes into both a neutron and a positron.

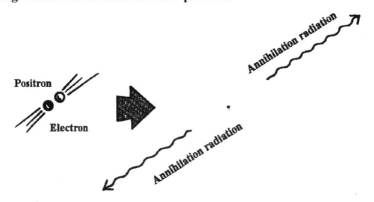

Fig. 11 Production of annihilation radiation: A positron and an electron meet and annihilate, and two waves (photons) of annihilation radiation are emitted 180 degrees apart.

---

* In contrast to β^+ decay, usual beta decay may be expressed as β^- decay, and, in contrast to β^+ rays, usual beta rays are described as β^- rays.

II. Scientific Basis

6-4. An orbital electron is sometimes captured by a nucleus.

When a proton in the nucleus captures an orbital electron and changes into a neutron, this type of decay is called **orbital electron capture (EC).** After an EC decay, a nucleus is left which has one proton less and one neutron more than the original nucleus (see Fig. 12).

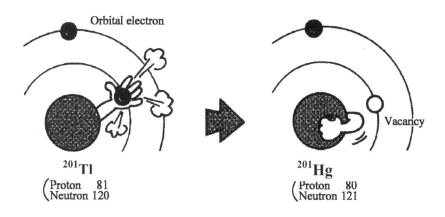

Fig. 12 Orbital electron capture in ^{201}Tl (thallium-201): A ^{201}Tl nucleus becomes a ^{201}Hg (mercury-201) nucleus after capturing one orbital electron. As this leaves a vacancy among the orbital electrons, emission of the characteristic X-rays of mercury follows. In orbital electron capture, one proton in the nucleus captures an orbital electron and changes to a neutron.

6-5. Gamma rays are electromagnetic waves emitted from an excited nucleus, and are essentially the same as X-rays.

A nucleus after decay may often be unstable, i.e., the nucleus is in an excited state. It becomes stable by emitting its extra energy in the form of electromagnetic waves, usually within a very short period of time (see Fig. 13). In this case, the electromagnetic waves emitted from the nucleus are called gamma (γ) rays. Such emissions do not cause the nucleus to change further into another type, and, for this reason, the term 'gamma decay' is not used.

Gamma rays and X-rays are both electromagnetic waves and are distinguished not by their energy (wave length), but by their origins. High energy electromagnetic waves (photons) other than gamma rays are categorized to be X-rays.

Sometimes, instead of emitting a gamma ray, an excited nucleus transfers its extra energy to one of its orbital electrons, and that electron is emitted, i.e. knocked out of the atom. This phenomenon is called "**internal conversion**", and the emitted electron is called an "**internal conversion electron**".

6-6. Gamma rays are sometimes emitted with delay.

Sometimes the excited nucleus is not extremely unstable but has a finite period of life. In such a case, the excited nucleus is said to be a "**nuclear isomer**", and is designated by adding 'm' at the end of its mass number, e.g., ^{99m}Tc (technetium-99m). Nuclear isomers have their own half-lives. They become stable by emitting gamma rays in a process known as "**isomeric transition**" (IT).

---

*Electromagnetic waves: X-rays/gamma rays are electromagnetic waves like radio waves, infrared rays, visible light and ultraviolet rays, but their wavelength is very short. Electromagnetic waves sometimes behave as particles. When electromagnetic waves are considered to be a stream of particles, the particles are called **photons**. The shorter the wavelengths are, the larger the photon energy is.

II. Scientific Basis

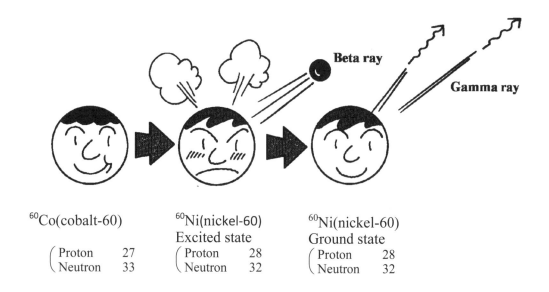

Fig. 13 Gamma ray emissions: Nuclei are left excited after decay, and become stable by emitting their extra energy as gamma rays.

6-7. Sources and action of neutrons with matter

As we know, neutrons are generally produced by fission of heavy nuclei such as ^{235}U (uranium-235) irradiated with low-energy neutrons, a familiar process occurring in the nuclear reactor.

There exist, however, some radioisotopes whose nucleus splits automatically into two fragments and at the same time emits several high-energy neutrons. This process is called the **"spontaneous fission"(SF)"**. The most widely used neutron source of such type is ^{252}Cf (californium-252).

Another type of neutron sources is those using nuclear reaction such as(α,n) reaction. The ^{241}Am-Be neutron source is an example.

The interaction of neutrons with other materials is very different from that of alpha and beta rays. Moreover, the action of neutrons varies greatly depending on their speed (kinetic energy). See Fig. 14.

Because the neutron has no electric charge, it can easily approach an atomic nucleus without electrically interacting with charged nuclei. Slow neutrons are readily absorbed by the nucleus. When a neutron is thus captured by a nucleus, the nucleus becomes an isotope of the same element with the mass number greater by 1. Such an isotope may often be radioactive. For example, stable ^{59}Co (cobalt-59) becomes radioactive ^{60}Co by capturing a neutron. Most radioisotopes are produced in nuclear reactors by using this process of **"neutron capture"**.

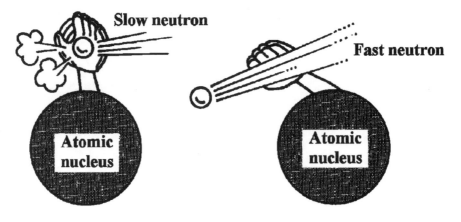

Fig. 14 Neutron capture by a nucleus: When a neutron is captured by an atomic nucleus, the nucleus becomes an isotope of the same element. Slow neutrons are captured more easily than fast neutrons.

II. Scientific Basis

Table 1 shows a comparison of the properties of alpha, beta, gamma (X-rays) and neutrons.

Table 1　Types of radiation and their properties

Properties	Alpha rays	Beta rays	Gamma rays (X-rays)	Neutrons
Nature	Helium nucleus	Electron Positron	Electromagnetic wave (photon)	Neutron
Mass	Large	Very small	None	Large
Electric charge	+ 2e	− 1e +1e	None	None
Penetration	Small	Medium	Large	Large
Photographic effect	Large	Medium	Small	Small
Fluorescence effect	Large	Medium	Small	Small
Ionization effect	Large	Medium	Small	Small

7. X-Rays
There are two types of X-rays: bremsstrahlung and characteristic X- rays.

X-rays are electromagnetic waves like visible light and radio waves. X-ray tubes are used to generate X-rays (see Fig. 15). Electrons emitted from a heated filament in a vacuum are accelerated toward an anode by high voltage. When they strike the metal anode, X-rays are generated. This is because, as the electrons are passing by atomic nuclei of the atoms of the anode, the electric force between the nuclei and the electrons exerts a strong braking action on the electrons, and their kinetic energy is partially converted to X-rays (see Fig. 16). X-rays generated in this way are called **bremsstrahlung** (a German term meaning braking radiation).

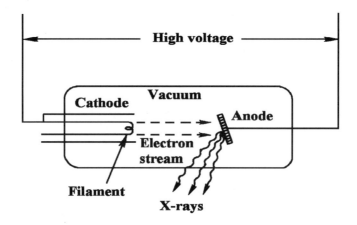

Fig. 15　X-ray tube:　Fast electrons collide with the metal anode and generate X-rays.

Additionally, when fast electrons cause the atoms of the anode to become excited or ionized, and electrons from outer orbitals move to vacant inner orbitals, energy equal to the energy difference between the outer and the inner orbit is emitted as electromagnetic waves (see Fig. 17). Such electromagnetic waves have energy specific to the kind of atom, and are called "**characteristic X-rays**" (see also Fig. 12). Instead of the emission of characteristic X-rays, sometimes one of the electrons in an outer orbit is given excitation energy and kicked out of the atom. This electron is called the "**Auger electron**" which is derived from French physicist Prof. Pierre Auger.

II. Scientific Basis

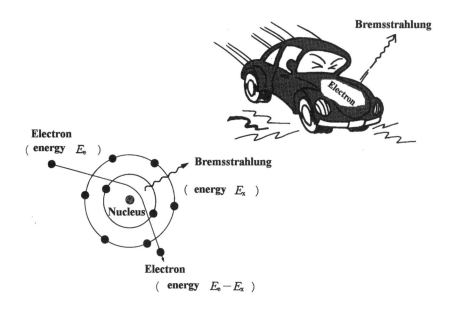

Fig. 16 Generation of bremsstrahlung: When moving electrons are braked, X-rays are generated.

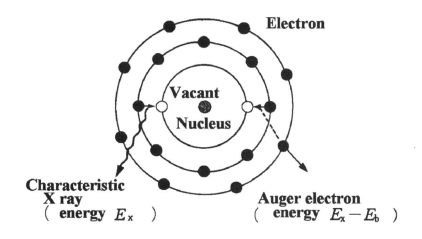

Fig. 17 Emission of characteristic X-rays or Auger electrons: When an electron in an outer orbit moves to a vacancy in an inner orbit, a characteristic X-ray or an Auger electron are emitted.

E_b: Energy necessary to release an orbital electron from the atom (binding energy of an orbital electron to the nucleus).

8. **Accelerators**

Accelerators are used to generate various kinds of radiation and to produce radioisotopes.

Because electrons and protons are electrically charged, they can be accelerated in a vacuum by application of electric force and given high kinetic energy. An accelerator is a device to produce fast streams of electrons, protons, or other charged particles. Various types of accelerators are used in academic studies, and for the production of high energy X-rays or radioisotopes for industrial and medical purposes. These accelerators are legally classified as radiation generators.

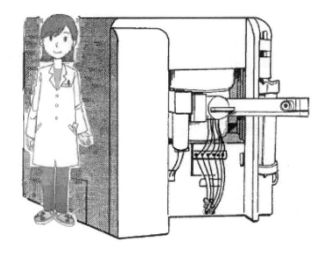

II. Scientific Basis

9. Types of Radiation

There are many types of what we commonly call "radiation".

As has been explained so far, radiation includes X-rays, alpha rays, beta-minus rays, beta-plus rays, gamma rays, neutrons, electrons, protons and cosmic rays. Major types of radiation can be classified by their characteristics as shown in the table below. Radiation that exists in the environment is called **"natural radiation"**. Radiation created artificially is called **"artificial radiation"**. **"Radioactivity"** and **"radiation"** are similar words and are often confused -- but the meanings are different. When a radioisotope is likened to an electric lamp, for example, radiation is the light emitted from the lamp, while radioactivity is the property of the lamp that emits light.

Radiation is the product of anything that is radioactive. It does not come only from radioisotopes, but also from accelerators and nuclear reactors, as well as cosmic rays.

Table 2   Major Types of Radiation

Radiation	Electromagnetic radiation (photons)	X-rays (Bremsstrahlung, characteristic X-rays, generated by phenomena occurring outside of nuclei)
		Gamma rays (generated by changes in the energy states of nuclei)
	Charged particles	Beta-minus rays (electrons emitted from nuclei)
		Beta-plus rays (positrons emitted from nuclei)
		Electrons (produced by accelerators)
		Alpha rays (helium nuclei emitted from nuclei)
		Protons (produced by accelerators)
		Deuterons (produced by accelerators)
		Various heavy ions and mesons (produced by accelerators)
	Uncharged particles	Neutrons (produced by reactors, accelerators and radioisotopes)

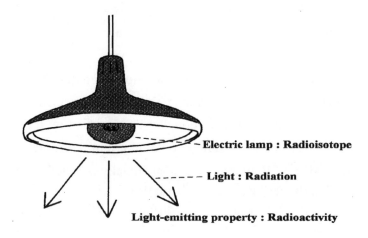

Fig. 19 Relationship among radioactivity, radiation and radioisotopes

II. Scientific Basis

10. Quantities and Units Related to Radiation

The basic unit of energy is the **joule (J)**; however, the usual measure of radiation energy is the **electron volt (eV)**. The kinetic energy gained by an electron when it is accelerated by a potential difference of 1 volt is defined as 1 eV. The relationship between the electron volt and the joule is,
  1 eV = 1.6022×10^{-19} J
  1 eV $\times 10^3$ and 10^6 are called 1 keV and 1 MeV, respectively. For example, ^{60}Co emits gamma rays of 1.17 MeV and 1.33 MeV after emitting beta rays with a maximum energy of 0.32 MeV.

In assessing the effect of radiation on humans, three quantities: absorbed dose, equivalent dose and effective dose, are considered. The equivalent dose and the effective dose are used for the ICRP's system for radiation protection and called the "**protection quantities**".

10-1. Absorbed dose

As a result of interaction between radiation and material (receptor), the energy absorbed per unit mass of the material is called the **absorbed dose**. The absorbed dose is a fundamental dosimetric quantity, and can be considered regardless of the kind of radiation or the kind of material. The unit for absorbed dose is the **gray (Gy)**. An absorbed dose of one gray means that one joule of energy is absorbed per one kilogram of the material.

When using the absorbed dose, it should be mentioned what the absorbing material is, because in a radiation field the absorbed dose will vary depending on the absorbing material. For radiation protection purposes, air, water or soft tissue are usually regarded as the absorbing material.

10-2. Equivalent dose

When the human body is exposed to radiation, the degree of biological effects will differ depending on the type and energy of the radiation, even in cases where the absorbed dose is the same. This is because the damage to DNA due to the energy deposited by radiation is more difficult to repair as the density of energy deposition becomes greater.

The concept of equivalent dose was created as a common index for calculations of risk to the human body from radiation under different conditions. This concept of equivalent dose is used only in the context of radiation protection calculations.

The relationship between equivalent dose and absorbed dose in a particular tissue or organ is described by the following formula:

Equivalent dose in Sv = Absorbed dose in Gy ×
radiation weighting factor (w_R)

The **radiation weighting factor** assumes that the degree of biological effect differs depending on the ionization density (**linear energy transfer, LET**) of radiation and determined by the ICRP based upon theoretical and experimental data and considering practical applications. Radiation weighting factors of 1 for beta rays (electrons) and gamma rays (X-rays), and 20 for alpha rays are currently used. For neutrons the value varies with their energy ranging 2.5 to about 20.

When the gray is used as the unit for absorbed dose, the unit for equivalent dose is the sievert (Sv). The millisievert (mSv, 10^{-3} Sv) and the microsievert (μSv, 10^{-6} Sv) are also used.

10-3. Tissue equivalent dose and effective dose

Equivalent dose for a particular tissue or organ T is called **tissue equivalent dose** H_T. When the human body is exposed to radiation, how the effects (in this case, incidence of fatal cancer or severe genetic diseases) appear depends on tissue or organ. In order to assess the total of such effects on various tissues and organs throughout the body, **effective dose** is used. To get the effective dose, tissue equivalent doses are multiplied by tissue weighting factors and then summed up for each exposed tissue and organ throughout the body. Tissue weighting factors (Table 9) are estimated from clinical data, followup studies of Atomic bomb survivors, and studies with experimental animals. The unit of effective dose is also Sv (See also IV 7 and 8).

II. Scientific Basis

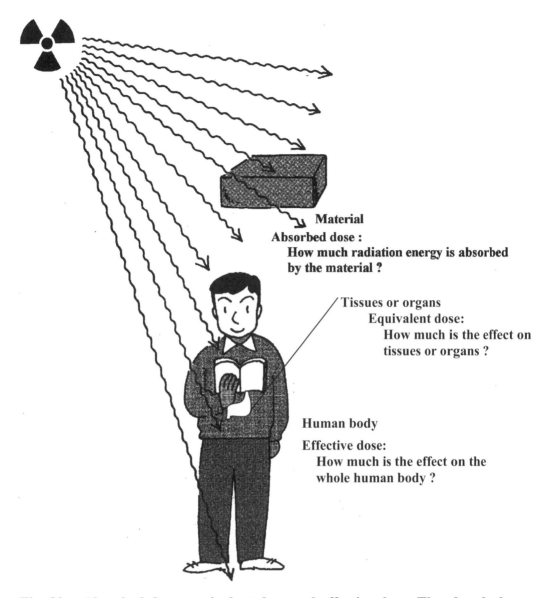

Fig. 20 Absorbed dose, equivalent dose and effective dose: The absorbed dose is a measure of how much energy is absorbed in material, regardless of the type of radiation or material; the equivalent dose is a measure of biological effects of radiation on tissues or organs in the human body. The effective dose is a measure of total effects on the whole body. The tissue equivalent dose and the effective dose are used solely in the context of radiation protection calculations.

10-4. Operational dose quantities and units

The tissue equivalent dose and the effective dose are almost impossible to be measured because it would require dosimeters inside of the body. Therefore, in practice, **operational dose quantities** are defined as substitutes for them for external irradiation. These operational dose quantities were provided by the International Commission on Radiation Units and Measurements (ICRU).

For the measurements of effective dose rate in a workplace, dose rate meters calibrated with the **ambient dose equivalent** (designated as **1-cm dose equivalent** in Japanese regulations) are used which is defined by the dose at a point of 1 cm depth in a sphere of 30 cm diameter made of tissue equivalent material (ICRU sphere).

For the measurements of skin equivalent dose in a workplace, dose rate meters calibrated with the **directional dose equivalent** (designated as **70μm dose equivalent**) are used.

For the measurement of personal dose, personal dosimeters calibrated with the **personal dose equivalent** are used.

The units of these dose equivalents are also Sv.

Table 3 Quantities and units related to radiation

Item	Unit	Symbol	Definition	Remarks
Absorbed dose	Gray	Gy	1 Gy corresponds to absorbed energy of one joule per kilogram of material	SI unit: J/kg
Equivalent dose	Sievert	Sv	Absorbed dose (Gy) × radiation weighting factor	SI unit: J/kg
Effective dose	Sievert	Sv	$\sum$(Equivalent dose for a tissue × tissue weighting factor)	SI unit: J/kg
Activity	Becquerel	Bq	One decay event per second	SI unit: s^{-1}
Radiation energy	Electron volt	eV	Kinetic energy gained by an electron when it is accelerated by a potential of one volt	SI unit: J

III. Safe Handling

1. Who is the Radiation Protection Staff?

Those who are in charge of the practice of radiation protection at an establishment are called the **radiation protection staff**. This is not a legally defined occupation, but no establishment can be operated without someone who is actually in charge of radiation protection. The number and positions of radiation protection staff will differ according to the size and nature of the establishment. The scope of work of radiation protection staff ranges from controlling the access of radiation workers to environmental monitoring, and to personal exposure control.

1-1. Responsibilities of radiation protection staffs and their relationships with radiation workers

The position within the organization, scope of work, and responsibilities of a radiation protection staff member will differ according to the establishment, and be defined in the Radiation Hazards Prevention Program of the establishment (see Chapter VI).

There may be a misunderstanding among radiation workers that the work of radiation control can be left in the hands of the radiation protection staff. In reality, actual radiation control cannot be done only by radiation protection supervisors (see Chapter V 7) and radiation protection staff, but requires the active participation of the radiation workers themselves.

1-2. Keeping close contact with radiation protection staff

Because the radiation protection staffs at each establishment are most knowledgeable about radiation control at that location, radiation workers should feel free to talk to him or her even about routine operations. Of course, whenever anything abnormal occurs − particularly any kind of accident − radiation workers must contact anyone of the radiation protection staff members promptly, and follow his or her instructions.

III. Safe Handling

Fig. 21 Radiation protection staff

2. Protection Against External Exposure

Measures for protection against external exposure should focus on the three principles: shielding, distance and time.

2-1. Shielding

Dose rate in any workplace should be reduced by shielding radiation sources with lead, iron or concrete for gamma rays, and plastics or water for neutrons. Specific measures will vary according to the type and energy of radiation. Shielding can generally be done more easily and more economically as dose to the radiation source as possible.

2-2. Distance

It is important to work as far as possible from the radiation source. If the radiation source is a point source, such as a gamma ray source, the dose rate is in inverse proportion to the square of the distance. For example, if the dose rate at a point one meter from the point source of gamma ray is 100 mSv/h, it will be one fourth, or 25 mSv/h, at a point two meters away, and only one hundredth, or 1 mSv/h, at a point ten meters away. The exposure can be substantially reduced by using remote handling equipment such as tongs.

2-3. Time

By shortening exposure time, the exposure can be reduced. Radiation workers should review operational procedures in advance, and work efficiently. Time reduction, however, should be looked to only after the best measures for shielding and distance have been taken.

III. Safe Handling

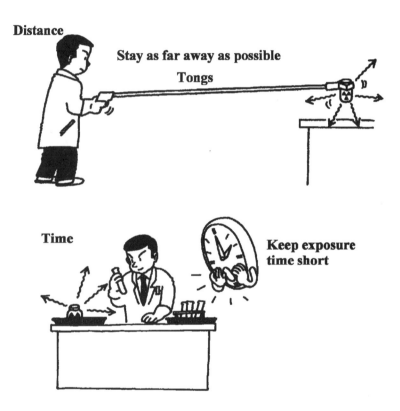

Fig. 22 Three principles for protection against external exposure: Those who handle radiation must always keep in mind the three principles of shielding, distance and time.

3. Protection Against Internal Exposure

Protection against internal exposure can be attained by faithfully observing all rules and guidelines for operations with unsealed radioisotopes.

Radioisotopes can enter the body through the following three intake pathway:
(1) through the respiratory system;
(2) through the digestive system; and
(3) through the skin, especially a wound.

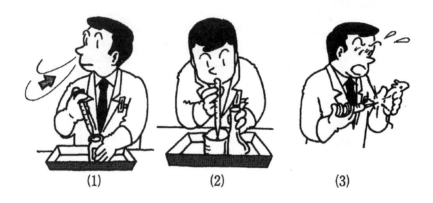

Fig. 23　Intake pathway of radioisotopes into the Body

Generally, intake through the respiratory system—inhalation—is the most significant. In order to prevent radioisotopes from entering the body, it is important to keep equipment and the workplace orderly, to work efficiently, and to faithfully observe all rules for operation.

In a facility where radioisotopes are handled, it is necessary to put on special working clothes. When leaving a controlled area, those clothes should be removed, hands should be well washed, and contamination on the hands, feet and clothes should be checked for by using a body surface monitor (a Hand-foot-clothes monitor). Whenever anything is brought out of the controlled area, it should be checked for radioactive contamination with appropriate measuring equipment. Continuous care must be taken in regard to radioactive air contamination and surface contamination on floors, tables, etc., in the working environment.

III. Safe Handling

When working in an area of possible air contamination, workers need to wear a protective mask to avoid inhalation of radioactive aerosol.

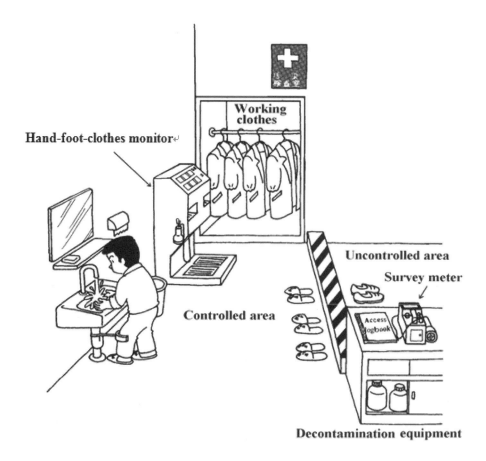

Fig. 24 In order to prevent radioactive contamination when handling unsealed radioisotopes, special working clothes, slippers and rubber gloves should be worn. When leaving the controlled area, working clothes should be removed, hands washed well, and contamination checked for by a body surface monitor.

When airborne dust, gas or vapor can be produced, unsealed radioisotopes should be handled under a hood (see Fig. 25). It is sometimes necessary to handle them in a fully airtight space, such as a glove box (see Fig. 26).

Eating, drinking, smoking and wearing make-up must not be done in a work room where radioisotopes are handled. Nothing should be sucked through a pipette by mouth.

In order to prevent radioisotopes from adhering to the body, special clothes, a cap, footwear and rubber gloves should be worn. If circumstances so require, a respirator or, sometimes, an airline suit should be worn.

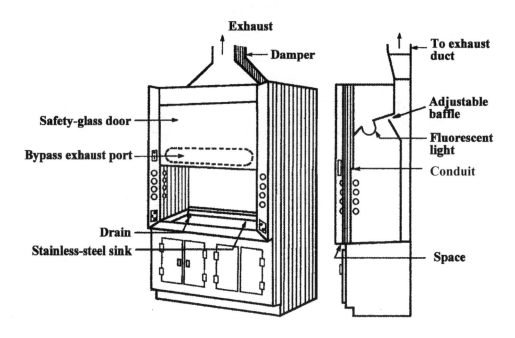

Fig. 25 Oak-Ridge-type hood

III. Safe Handling

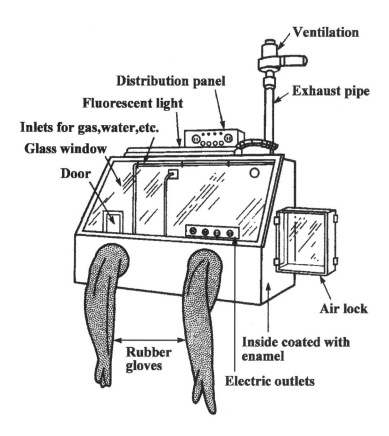

Fig. 26　Glove box

4. Attitude when Handling Radioisotopes and Radiation

When handling radioisotopes and radiation, for protection against both external and internal exposure, the basic three-point attitude is: "Don't be afraid; don't panic; but don't take it lightly." For that, it is necessary to have an accurate understanding of radioisotopes and radiation.

In addition, when commencing a new task, it is of course necessary to plan the work carefully. It is also of great benefit in the prevention of accidents to practice the necessary operations several times without using real radioisotopes or radiation (cold run), in order to become familiar with the steps and procedures, and to modify them if necessary.

Fig. 27 Attitude when handling isotopes and radiation

III. Safe Handling

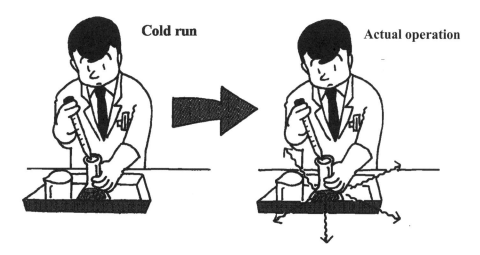

Fig. 28 Working without a rehearsal could cause an accident

5. Radiation Monitoring

Radiation cannot be felt or detected by our senses, but its existence and amount can be determined with appropriate methods.

5-1. Personal monitoring instruments

(1) External exposure

Thermoluminescent dosimeters (TLD), fluoroglass dosimeters(FLD), optically stimulated luminescence (OSL) dosimeters and electronic personal dosimeters are all instruments to measure personal dose equivalent from external exposure (see Fig. 29). These dosimeters are worn on the chest (or on the abdomen for women of reproductive age). If there is any possibility of substantial exposure to other parts of the body, for example, the fingertips, those have to be monitored with special monitors such as ring badges.

(2) Internal exposure

To assess doses from internal exposure, the kind of radioisotopes and their amount of intake must be known. There are several methods for determining the intake of an isotope. Intake can be determined externally by means of a whole-body counter (Fig. 30), by measuring activity concentration in the air at workplace and working time, or by the method of bioassay, i.e. radiochemical analysis of excreta. The internal dose can then be calculated based on radioisotope intake as obtained by any of these methods.

When using a whole-body counter, the amount and distribution of gamma rays can be determined by measuring gamma rays emitted by the radioisotope distributed in the body from outside. In contrast to this, in the bioassay method, radioisotope contained in urine, feces or, in cases of handling tritium, breath is used to estimate the amount of radioisotope in the body. The bioassay is appropriate to assess the intake of radioisotopes emitting only beta rays or alpha rays, but it takes much effort and time.

III. Safe Handling

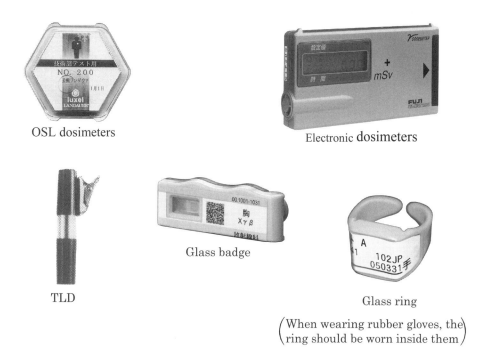

Fig. 29 There are a variety of instruments to measure personal external exposure, which are used separately or jointly.

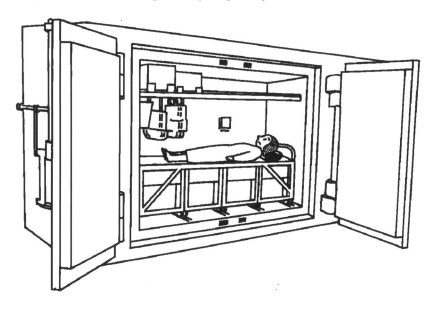

Fig. 30 Whole-body counter: This device is sensitive to gamma rays enough to easily detect even the slight amount of potassium (^{40}K) that is naturally present in the body.

5-2. Workplace monitoring

Using survey meters, it is easy to measure the dose rate in working areas. Geiger-Mueller (GM) counters, ionization chambers and scintillation counters are common types of survey meters, and each have their own strengths and weaknesses (see Fig. 31).

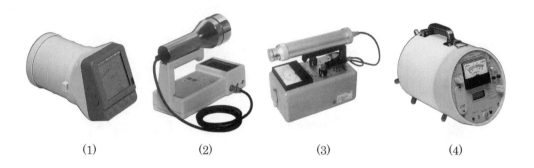

(1)　　　　　(2)　　　　　(3)　　　　　(4)

Fig. 31　Major types of survey meters
   (1)　Ionization chamber type
   (2)　GM counter type
   (3)　Scintillation counter type
   (4)　Neutron survey meter

In order to measure activity concentration in the air or in water, various kinds of sampling equipment and radiation measuring instruments are used, in accordance with the specific needs of that measurement.

Radiation monitors, such as area monitors, water monitors and gas monitors, are placed at specified locations to continuously measure and monitor the amount of activity concentration at that spot.

III. Safe Handling

5-3. Monitoring of surface contamination

Survey meters can sometimes be used to check for surface contamination on a floor, tables and any other goods, but when the dose rate in the vicinity is high, e.g., in the case of a container of gamma ray sources, contamination on the surface can often not be measured accurately. In such cases, wipe tests are used. In that method, extraneous matter on the tested surface is wiped off with filter paper (see Fig. 32). Radioactivity on the paper is then measured to detect and assess surface contamination.

For contamination on workers' bodies, body surface monitors are used at the exit of controlled areas.

Fig. 32 Collecting a sample for a wipe test

5-4. External exposure measurement

External exposure refers to exposure when the radiation source is outside the human body. Exposure levels depend on the type and energy of the radiation. Low-penetrating radiations such as alpha rays and low-energy beta rays are blocked from reaching the body by the air, by clothes, or by the surface layer of the skin. For low-penetrating radiation, the dose equivalent at a depth of 70μm is measured and used to assess the skin equivalent dose. Otherwise, 1-cm dose equivalent must be measured to assess the effective dose.

(1) Methods for measuring external exposure

External exposure can be assessed either by measuring the radiation level in the working environment or by measuring the exposure of an individual radiation worker. Measurement and assessment of individual dose are done continuously by personal dosimeters (Fig. 29) while working in the controlled area. When the whole body is evenly exposed, the instruments should be worn on the chests for men, while on the abdomen for women.

(2) When exposure is uneven

Sometimes, because of uneven distribution of radiation, only a part of the body receives significant exposure to radiation, for example, in the case of exposure to X-rays or beta rays from small sources, or by wearing a protective apron. In such situations, tissue equivalent doses should be considered for the three regions of the body shown in Fig. 33.

(3) When measuring dose is difficult

It is sometimes difficult to measure and assess dose with personal dosimeters. In such a situation, a survey meter or other radiation measuring instrument is used. If this, too, is difficult, assessment based on calculations is permissible.

(4) Surface contamination on the skin and clothesing

Usually hand-foot-clothes monitors are used.

III. Safe Handling

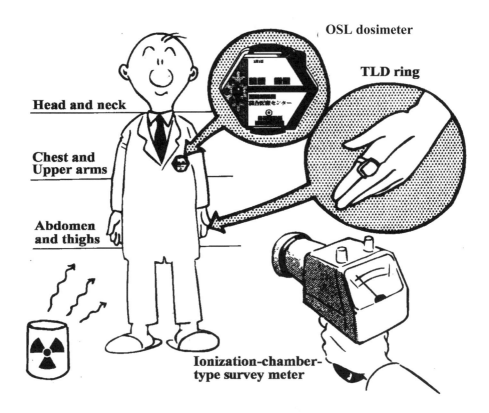

Fig. 33　Measuring external exposure

5-5. Internal exposure and its assessments

Internal exposure refers to the exposure when radiation sources are inside the body, i.e. when any radioisotope has somehow entered the body. With internal exposure, unlike with external exposure, exposure to alpha rays or low-energy beta rays becomes significant.

(1) Pathway for internal exposure

Radioisotopes can enter the body by inhalation, by ingestion, or through the skin, especially through a wound, all of which are referred to as intake. Radioisotopes move through the body along different pathway, depending on their physical and chemical characteristics, and are eventually excreted from the body by urine and feces, and sometimes by exhalation.

(2) Methods for assessing internal exposure

The intake of radioisotopes can be assessed by external measurement of gamma rays emitted by the radioisotopes within the body. For this, a whole-body counter is used.

The bioassay method indirectly measures the intake of radioisotopes by measuring their presence in samples of excreta, such as urine and feces. For this, biokinetic models are necessary, which allow assessments of the amount of radioisotopes deposited in various organs. This method is useful in analyzing radioisotopes that emit only alpha or beta rays.

Calculations based on measured concentrations of radioisotopes in the air involve many variables, including the amount of inhalation, breathing rate, and how long one stays in the place. Despite these difficulties, the method is widely used.

(3) Implementation of the assessments

The calculation of internal exposure is done for an individual who may take in radioisotopes inadvertently. In actual practice, the appropriate method or

III. Safe Handling

combination of the methods is determined based on a judgment of cost, labour, the kind of radioisotopes handled, the nature of the work operation, and the working environment. When the inhaled or ingested amount becomes known, assessment of effective dose or tissue equivalent dose is done by using dose-to-intake conversion coefficients given by the ICRP. These coefficients are also given in the Notification by the **NRA**. The Act on Prevention of Radiation Hazards due to Radioisotopes, etc. requires determination of internal exposure at least once every three months for those workers who work in the area where radioisotope intake may occur.

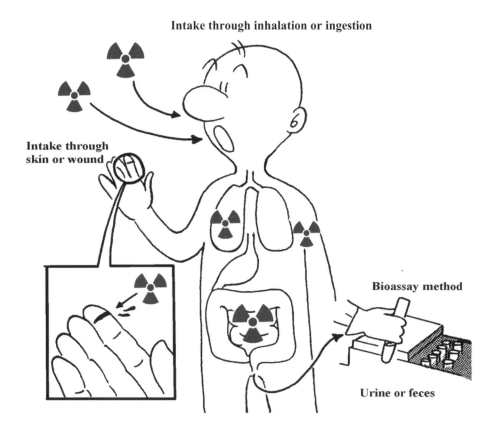

Fig. 34   **Measuring internal exposure**

6. Education and Training

It is essential for radiation protection that radiation workers be aware of radiation-control requirements, and that they be fully cooperative. To this end, radiation workers are required, through education and training, to understand the importance of radiation control and specific protection measures, and to have basic knowledge about biological effects of radiation. Education and training are necessary before a radiation worker deals with radiation or radioisotopes for the first time, and regularly thereafter at least once a year.

6-1. Timing of education and training

Radiation workers should receive education and training (1) prior to their first entering to controlled areas; and (2) at least once every year thereafter (re-education). Through re-education, radiation workers obtain the latest knowledge on radiation effects and techniques for safe handling, as well as have the opportunity to reconfirm regularly the importance of safe handling.

6-2. Topics to be learned through education and training

For inexperienced radiation workers, items to be taught and required hours are clearly established, as shown in the Table 4. These are minimum requirements, with each establishment providing its own education and training curriculum, taking its own situation into account. Programs often go beyond mere lectures, and include opportunities to improve the skills of safe handling in realistic situations. It is also important that radiation workers learn from their more experienced colleagues and from radiation protection staffs — which is to say, on-the-job training, or OJT.

III. Safe Handling

Table 4   Required items and length of education/training before entering a controlled area or beginning operations

Item / Classification	A	B
Effects of radiation on human body	30 min	30 min
Safe handling of radioisotopes and radiation generators	4 hrs	1.5 hrs
The Act and relevant Ordinances on prevention of radiation hazards due to radioisotopes, etc.	1 hr	30 min
Radiation Hazards Prevention Program	30 min	30 min

A: Radiation workers
B: Workers who do not enter controlled areas.

Fig. 35   Education and training

7. From Procurement to Weste management of Radioisotopes

At any establishment dealing with radioisotopes, there are various steps from procurement of radioisotopes to pre-disposal of them. Vital parts of each of these steps are safe handling, control and recording. Especially, the types and amounts of sealed radiation sources being brought in and taken out must be balanced. To accomplish this, at each step — procurement, storage, use, transportation, and disposal — safety control and accurate record-keeping are essential.

7-1. Procurement

Only permitted radioisotopes and their quantities for which there has been proper notification may be received. Usually the type and quantity of the radioisotopes that can be used in each workplace are limited, so it is necessary to talk with the radiation protection supervisor and radiation protection staffs before receiving them.

It is also important that more radioisotopes than necessary are not purchased, as this will only increase the management burden.

7-2. Storage

When the radioisotopes are delivered, they are usually put into a storage facility, where they are recorded in a logbook.

7-3. Use

Prior to actual use, it is recommended that a cold run be undertaken in exactly the same way as the operation to be performed. An amount as close as possible to the exact amount required should then be taken out of storage, and any leftover should be returned to storage, with the appropriate recording entries made.

7-4. Transportation

When radioisotopes (including materials contaminated by radioisotopes) are transferred outside the facility, the type and quantity should be recorded, and they should be transported in accordance with the requirements for transportation

III. Safe Handling

standards. No one should bring any radioisotope out of a facility without instructions or the approval of the radiation protection supervisor.

7-5. Waste management

Solid and liquid radioactive waste produced as a result of usage, etc., should be sorted according to the categories, Table 5, in the room where the radioisotopes are being used, and then transferred to a radioactive waste-storage facility. The waste should be classified accurately into the prescribed categories.

7-6. Restrictions on transfer, procurement and possession of radioisotopes

Permitted users are not allowed to transfer any radioisotopes beyond what is stated in the permit to any other user, seller or waste management operator, nor to accept or hold radioisotopes in excess of the capacity of the facility as mentioned in the permit.

8. Usage of Radioisotopes and Radiation Generators

The application of radioisotopes and radiation generators varies significantly depending on the nature of the radiation source. Radiation sources can be broadly classified as sealed sources, unsealed sources, and radiation generators. Proper handling and radiation control also differ according to the sources.

8-1. Use of sealed sources

The followings are the principal advantages of using sealed sources:
(1) There is no risk of leakage or damage during use under normal conditions.
(2) There is no risk of the radioisotopes being scattered and causing contamination as a result of leakage, etc. Sealed radiation sources, however, do have the potential to cause radioactive contamination if there is a problem with the sealing. To avoid this, it is necessary to check periodically for contamination.

8-2. Use of unsealed sources

Unsealed radiation sources should be handled only in a 'working room'. The inside walls and floors of the working room, which may become contaminated, must be constructed with a minimum of projections, indentations, cavities and joints. Materials should be flat and smooth, and resistant to permeation and corrosion. The radiation source itself should be handled in a hood or a glove box.

8-3. Use of radiation generators

In case where induced radioactivity is not produced by the use of radiation generators, it is sufficient to control external exposure in the same manner as for sealed radiation sources. On the other hand, if an activation product is generated, controls similar to those for an unsealed radiation source are sometimes required.

III. Safe Handling

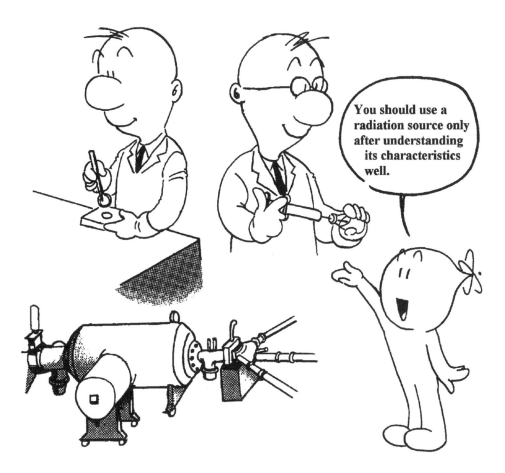

Fig. 36 Usage of radioisotopes and radiation

9. Storage of Radioisotopes

When radioisotopes are not being used, usually they should be put in an appropriate container and stored in a fireproof storage room. Sealed sources, however, can be stored in fire-proof containers in a storage facility.

9-1. Requirements for storage facilities

The main structure of the storage facility has to be of fireproof construction, and radioisotopes beyond the approved storage capacity should not be stored there. Radioisotopes should be put into containers and stored in a fireproof storage room or storage box, in such a way that they cannot be taken out easily without permission. Special fire doors should be used at the storage room entrance, and locks and other security equipment should be installed.

9-2. Requirements for storage containers

Containers for radioisotopes have to meet the following requirements:
1) They should be air-tight so as to prevent contamination of the outside air.
2) Containers for liquid radioisotopes have to be suitably impermeable and designed so that spilling is avoidable.
3) If there is any danger that a container for liquid or solid radioisotopes might become cracked or broken, it should be used with a saucer, absorbing material or other means to absorb and prevent the spread of the radioisotope.

9-3. Requirements for waste-storage facilities

The above requirement are also applicable to the technical requirements for waste-storage facilities.

III. Safe Handling

Fig. 37 Storage of radioisotopes

10. Discharge of radioisotopes from facilities and radioactive waste management

The Radiation Hazard prevention Act and its Ordinances regulate the discharge of radioisotopes from facilities and radioactive waste management.

10-1. Procedures for discharging radioactive materials in exhaust and drainage

Radioactive materials in exhaust are mostly discharged into the atmosphere within the radioactivity concentration not exceeding the limit specified by the NRA, through an exhaust (or ventilation) system equipped with a high-efficiency filtering unit to remove radioactive particulate.

Drainage with lower radioactive concentration are collected in a reservoir, diluted with non-radioactive drain water and discharged with the radioactivity concentration not exceeding the limit specified by the NRA to the public sewage system.

10-2. Radioactive waste management (Pre-disposal of liquid and solid wastes)

Liquid materials of higher radioactive concentration are treated by the several methods of processing; e.g. 1) after evaporation or coprecipitation to reduce radioactive concentration, and then the liquid phase is discharged to the public sewage system, the residue being treated as solid waste, 2) after being sealed in a container, which is solidified with solidification materials and stored for disposal in a waste-storage facility, 3) are incinerated to reduce volume, the residue being treated as solid waste.

Solid radioactive wastes have to be stored in waste-storage facilities until a waste management operator collects the wastes. Solid wastes should be divided into several categories and put into separate containers for the convenience of which method of pretreatment will be used afterwards.

10-3. An example of the categorization of Radioactive wastes

An example of the categorization of solid wastes is shown in Table 5. These categories apply to the collection of wastes from radioisotope users by the Japan Radioisotope Association.

III. Safe Handling

Table 5 An example of waste categories for separate collection by the Japan Radioisotope Association

Classification	Items
Combustible Type I	Paper, clothes, wood pieces
Combustible Type II	Plastic tubes, plastic vials, polyethylene sheets, rubber gloves
Incombustible (Compressible)	Glass vials, other glass equipment, injection needles, vinyl chloride pipes, vinyl chloride
Incombustible (Incompressible)	Soil, sand, iron bars, pipes, concrete pieces, castings, clock parts, large amounts of TLC plate, machinery and equipment
Animal carcasses	Animals after being dried
Inorganic liquids※1	Fluids after testing
Combustible filters※2	HEPA filters, pre-filters, charcoal filters

Containers are normally 50-liter drums.

※1. Inorganic liquids are put first in 25-liter polyethylene bottles, which are then put into 50-liter drums. Adjust the PH to between 2 and 12, do not use HCl to adjust PH.

※2. Filters are packed in corrugated cardboard, polyethylene-sheets.

11. Record Keeping (Radioisotopes tracking)

The amounts of radioisotopes used and waste management of should be accurately entered in a logbook. The purpose of such entries is to ensure the management of radioisotopes facilities, by recording each step in the process from procurement of radioisotopes to radioactive waste managernent. The records must be maintained for a prescribed period of time.

11-1. Entry items

Table 6 below shows the items that are legally required to be entered. Entries make it possible to confirm the location of radioisotopes and how they are being handled at each step in the process from procurement of radioisotopes to radioactive waste management, for all radioisotopes at the establishment. Entry is a very important task from the viewpoint of radiation source control at an establishment.

Table 6　Items to be entered by radioisotope or radiation generator users

Entry Items
1.　Use of radioisotopes
2.　Operation of radiation generators
3.　Storage of radioisotopes
4.　Transportation of radioisotopes
5.　Radioactive waste management
6.　Inspection of radiation facilities
7.　Education and training

11-2. Who should make entries?

There are two kinds of entries — those made by radiation protection staff and those made by radiation workers. Radiation workers, in principle, record the use and storage of radioisotopes, as well as transfer of radioisotopes and materials contaminated by radioisotopes. Because information relating to radioisotopes, from procurement of radioisotopes to radioactive waste management, amounts to a kind of "registry" for the radioisotope, record

III. Safe Handling

keeping should be done accurately. The format of the log book may be decided by each establishment, so as to make its use easy for the radiation workers, taking into consideration the nature of the work in the establishment. To ensure accuracy, an entry should be made every time a radioisotope is used, stored or transferred.

Accurate and timely entries should be made, not only during periods of normal operation, but when accidents or emergencies occur, updated regularly.

11-3. Custody period for entries

Log books should be closed annually, and then be maintained for five years.

Fig. 38   Record Keeping

12. Record Keeping (Dose)

Records should be kept on the results of all legally required measurements. Such measurements relate to the working environment and to personal exposure. Recording is thus for the purpose of controlling personal exposure.

12-1. Items to be measured

Items legally required to be measured are: 1) For places where there is a probable radiation hazard, the level of radiation and the state of contamination from radioisotopes (measurement of the working environment). 2) For those who enter radiation facilities, the dose received and state of contamination by radioisotopes (assessment of personal dose).

12-2. Measurements and recording of external exposure

External dose must be measured, and the results recorded, for the three-month periods starting on each April 1, July 1, October 1 and January 1, and for one-year periods starting each April 1. For pregnant women, measurements must be made and recorded every month. Items to be recorded are: 1) name of worker measured, 2) name of person doing the measurement, 3) type of the personal dosimeter used, 4) measurement method, and 5) body parts measured and results of the measurements.

12-3. Measurements and recording of internal exposure

Internal dose to workers who enter places where there is a possibility of internal exposure must be assessed at least once every three months (at least once every month for pregnant women), and the results of those assessments must be recorded each time. Items to be recorded are: 1) date of assessment, 2) name of worker being measured, 3) name of person doing the assessment, 4) kind and type of radiation measuring equipment, 5) assessment method and, 6) results.

III. Safe Handling

12-4. Custody Period for Records

Records of measurements of the working environment must be kept for five years. Records on the exposure and medical examinations of radiation workers must be kept permanently. It is recommended that records of accidents and emergencies also be kept permanently.

Fig. 39 Recording

13. Medical Examinations

Those who are to be assigned to handle radiation and radioisotopes must be checked to see if they are suitable for the job from the viewpoint of their health. The radiological medical examination is in contrast to an ordinary medical examination given to every worker in the establishment. There are two kinds of radiological medical examinations—one prior to starting work as a radiation worker and those given periodically thereafter.

13-1. Why is the medical examination necessary?

Each organization conducts two kinds of medical examinations—routine medical examinations for every worker, and radiological medical examinations exclusively for radiation workers. The purposes in both cases are not only to improve the health of the workers and allow them to work without anxiety, but to discover any health-related abnormalities as early as possible so that they can be treated properly.

13-2. Examination items legally required

Radiological medical examinations consist of oral interviews, examinations and tests. Examination and test items are shown in the table below. These were determined from past experience and are based on a knowledge of radiation hazards and radiation effects.

	Test and Examination Items
1)	Amount of hemoglobin, the counts of red and white blood cells and the percentage of each subtype of white blood cells in the peripheral blood
2)	Skin
3)	Eyes
4)	Any other parts of the body, or items, designated by the Nuclear Regulation Authority

III. Safe Handling

13-3. When are medical examinations given?

According to the Regulations on Prevention of Ionizing Radiation Hazards under the Labour Standards Law, radiological medical examinations should be conducted before starting the handling of the worker's first handling of radiation or radioisotopes, and every six months thereafter. If certain required conditions are met, however, radiation medical examinations can be omitted. On the other hand, whenever there has been any abnormal radiation exposure (exposure to radiation in excess of the effective dose limit) or abnormal contamination, workers must be examined by a doctor.

13-4. How to interpret the results of medical examinations?

If radiation and radioisotopes are controlled according to all applicable rules, radiation workers will not have any related health problems. Moreover, because the dose to each radiation worker is controlled by personal monitoring, there is practically no chance that a worker will be found in an examination to exhibit any abnormal values as a result of his or her normal radiation work.

Results of blood counts, one of the items in a radiological medical examination, commonly fluctuate for a variety of reasons. Thus, if an abnormal value is found, it is entirely possible that the body has a problem due to a cause other than radiation. The health of radiation workers should be evaluated based on the results of both radiological medical examinations and general medical tests.

14. Procedures at the Time of an Accident or Emergency

14-1. What is an accident and what is an emergency?

When radioisotopes are stolen or missing, the situation is referred to as an accident. When there is a danger of a radiation hazard, or when it actually occurs, the situation is referred to as an emergency.

14-2. Principles for immediate measures at the time of an accident

When the whereabouts of a radioactive source becomes unknown, the worker should make contact with a radiation protection staff member and ask for instructions on measures to be taken.

14-3. Principles for immediate measures at the time of an emergency.

Immediate measures to be taken by a radiation worker at the time of an emergency are: 1) avoid endangering him- or herself, or others, and ensure safety, 2) promptly notify others about the occurrence of the emergency, and 3) take measures to prevent contamination from spreading and to prevent the situation from worsening. It is necessary for radiation workers to learn about possible emergency situations in detail, through education and training, so that they will be able to accurately assess them and implement all needed measures immediately and effectively.

14-4. To whom and how to notify?

It is essential for radiation workers to immediately let the people around them know that an accident has occurred, in order to secure their safety and to prevent the spread of contamination. They should also notify a radiation protection staffmember of the accident and ask for instructions. Procedures for notification are posted at various locations where radiation workers can easily see them, such as at the entrances to controlled areas.

When notifying, the radiation workers should calmly state 1) when, 2) where, 3) what, and 4) why. When a fire occurs, the radiation protection supervisor or the operator is required to contact the fire department, and to report

III. Safe Handling

to the **NRA**.

14-5. Specific measures to prevent worsening of an accident

The way to prevent an accident from worsening depends on the situation. In the event of contamination with unsealed sources, identifying and clearly marking off the contaminated area is the key to preventing contamination from spreading.

14-6. Preventing reoccurrence of an accident

In order to prevent the reoccurrence of an accident, it must be analyzed carefully to accurately determine its cause. Then, anything needed to improve the facility, equipment or method of handling should be done as promptly as possible. For the investigation to be effective, the cause of the accident and the responsibility for it should be pursued separately.

14-7. Earthquake

In the case of a facility experiencing an earthquake with seismic intensity of 4 or larger on the Japan Meteorological Agency seismic intensity scale (sihndo 4 or larger) a radiation protection staff member of specified permitted user must contact the NRA.

15. Procedures in the Event of Excess Exposure or Contamination

If excess exposure or contamination occurs, radiation workers should immediately report to the radiation protection staff to ask for appropriate instructions. When an accident does occur, however, avoiding danger to the body and preventing spread of contamination are the first priority.

15-1. What is excess exposure?

When exposure exceeds the dose limit for occupational exposure, it is called excess exposure. Such a dose limit is not the limit of safety, however, and a radiation hazard does not necessarily exist even at excess exposure.

15-2. Measures in the event of excess exposure

If there is any possibility that a radiation hazard involving excess exposure exists, it is necessary to see a doctor without delay. The cause of the accident should then be investigated to avoid reoccurrence.

15-3. What is contamination?

Contamination refers to the existence of radioisotopes where they should not be. Contamination can occur on floors, table surfaces where radioisotopes are being handled, or on the surface or inside of the body. Radiation workers should make it a routine to check their workplace frequently with, for example, a survey meter, both before and after handling of radioisotopes, in order to detect contamination. Radiation workers should also check for contamination with a hand-foot-clothes monitor in a contamination monitoring room when leaving a controlled area.

15-4. Measures in the event of contamination

In order to prevent contamination from spreading, decontamination should be carried out immediately, after clearly identifying the contaminated area. Radiation workers have to notify a radiation protection staff member of the contamination and ask for instructions on the most suitable method of decontamination under the circumstances.

III. Safe Handling

15-5. Transfer of a radiation worker who received excess exposure

Whenever necessary, transferring a radiation worker who experienced excess exposure to a different department, or restricting his or her access to the controlled area, should be considered.

Fig. 40 Notification

IV. Biological Effects of Radiation

1. Classification of Radiation Effects on the Human Body

Effects of radiation on the human body can be categorized into those that appear in the person who are exposed and those that appear in his / her offspring. The former are called **somatic effects**; the latter are called **hereditary(inherited) effects**.

Somatic effects are further categorized into **acute effects** and **late effects**, based on the time from radiation exposure to the appearance of the effect. Acute effects appear within a relatively short time after exposure (usually in the several weeks). Late effects appear much later several years to decades. The period until the effect appears is called the **latent period**.

Even when total dose is the same, effects can be different depending on whether the exposure was all at one time (**acute exposure**), or little by little on multiple occasions or over a long period of time (**chronic or protracted exposure**). This is because, in the latter case, the body is able to exercise its inherent ability of recovery between the instances of exposure. Effects also differ between **whole-body exposure** and **partial-body exposure** (see Fig. 41).

Fig.41 Whole-body exposure and partial-body exposure: With whole-body exposure, concern is for the effect on parts of the body that are radiosensitive. With partial-body exposure, concern is only for the parts exposed to the radiation.

IV. Biological Effects of Radiation

2. Acute Effects

In the case of a high dose received all at one time, symptoms of acute effects differ depending on the part of the body exposed and the dose.

As an example of acute effects in the case of whole-body exposure, single exposure to gamma rays (X-rays) will be explained here. Symptoms differ among individuals, but typical cases are shown in Table 7. No clinical symptoms are recognized at doses of 0.25 Gy or less. A dose of 0.5 Gy causes a temporary reduction in the number of white blood cells, but the cell count returns to normal after a while. Radiation sickness results from an exposure of 1.5 Gy or more, and includes symptoms similar to those of a hangover. A dose of about 4 Gy brings death within 60 days to 50% of those exposed without appropriate medical care. At 7 Gy, the probability of death is almost 100%.

Table 7 Symptoms of acute effects and dose delivered
(Whole-body, single exposure to gamma rays (or X-rays))

Dose (Gy)	Symptoms
0.25 or less	Almost no clinical symptoms
0.5	Temporary reduction of white blood cells (lymphocytes)
1	Nausea, vomiting, whole-body languor, substantial reduction of lymphocytes
1～2	Death to 10% within 60 days ∗
～4	Death to 50% within 60 days ∗
5～7	Death to 90% within 60 days ∗
7～10	Death to 50% within 2 weeks

∗) without medical care (aseptic treatment, bonemarrow transplantation etc.)

In this table, absorbed dose (in Gy) is used instead of equivalent dose or effective dose (in Sv). That is because the radiation weighting factors (see 7-2) recommended by the ICRP are based on the late effects such as cancer induction and not applicable to the acute effects mentioned here.

3. **Late Effects**

With late effects, the latent period can be up to decades. Typical late effects are cancer and cataracts.

The latent period for cancer varies according to the exposed organ or tissue, age at the time of exposure and dose, but typically ranges from several years to decades. The probability of cancer incidence varies depending on the organ or tissue and age, but, according to the International Commission on Radiological Protection (ICRP), the incidence of fatal cancer in the case of whole body exposure to 10 mSv is 5 in 10,000 for workers.

The latent period for cataracts — the clouding (opacification) of the lens of the eye — ranges from several years to a few decades, varying with dose. The minimum dose needed to cause the incidence of an effect is called the threshold dose*. The threshold dose for cataract is considered to be 0.5 Gy.

Conditions that is caused by radiation exposure can also be caused by other causes unrelated to radiation. Thus, it is difficult to distinguish whether cancer or cataracts are the result of radiation exposure or not.

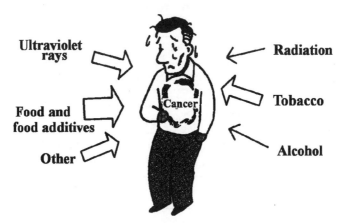

Fig. 42 Today in Japan, cancer is the leading cause of death. Various factors are involved in cancer incidence, and radiation is only one of them. It is not possible to distinguish cancer caused by radiation from cancer arising from other causes.

---

* Strictly speaking, the term 'threshold value' refers to the dose at which a specified effect occurs in 1 percent of those who have been exposed (see Fig. 43).

IV.　Biological Effects of Radiation

4.　Heritable Effects(Genetic Effects)

Heritable effects caused by radiation have not been verified in human beings.

The heritable effects of radiation have been studied for a long time, and no instances of their occurring in human beings have yet been confirmed even in children of atomic bomb survivors in Japan. Recently, the United Nations Scientific Committee on the Effects of Atomic Radiations (UNSCEAR) studied this problem by using the method of doubling dose (the amount of radiation required to produce the same number of mutations as that occur spontaneously in one generation of the population).

Because there are no data on radiation-induced mutations in humans, the doubling dose has been estimated using spontaneous mutation rates of human genes and radiation-induced mutation rates of mouse genes for low-dose, sparsely ionizing radiation of the order of one gray in this study.

For a population exposed to radiation in one generation only, the risks to the progeny of the first post-radiation generation were estimated to be 3000 to 4700 cases per gray per one million progeny. This constitutes 0.4 to 0.6 percent of the baseline frequency of those disorders in the human population as compared with the total risks (somatic and heritable) of approximately 6 percent per gray as described in the 1990 Recommendations of the ICRP (UNSCEAR 2000).

5. Biological Effects Depend on the Dose Received

Radiation effects on the human body can also be categorized into stochastic effects and deterministic effects, depending on the relationship between the radiation dose and the rate of occurrence.

Threshold values, as shown in Fig. 43 (left graphs) have been determined for acute effects (erythema on the skin, loss of hair, reduction of white bloods cells, etc.) and cataracts, one of the late effects. Beyond the threshold dose of exposure, the larger the dose is, the greater the rate of occurrence, and the greater the degree of effect (severity) will be. Such this kind of effects are called **deterministic effects**.

In contrast, occurrences of cancer including leukemia as well as heritable effects simply increase with the dose delivered as shown in Fig. 43 (right graphs). No threshold values are assumed to exist, and the severity of the effect has nothing to do with the dose. Such kinds of effects are called **stochastic effects**.

The various kinds of radiation effects on the body described above can be classified as shown in Table 8.

The relative contribution of a specified organ or tissue (T) to the total stochastic effect, as a result of even irradiation of the whole body, is called the tissue weighting factor (w_T) for that organ or tissue. **Effective dose** (E) is defined by the following formula:

$$E = \sum_T w_T \cdot H_T$$

Here, w_T is the tissue weighting factor for an organ or tissue T, and H_T is the equivalent dose for the organ or tissue T (see also Table 9).

Within the range of low doses involved with properly implemented radiation protection, the chances of a worker suffering from a deterministic radiation effect are virtually zero. There must be serious concern, however, about stochastic effects. Thus, measurements and evaluation of effective dose are very important in terms of radiation protection.

IV. Biological Effects of Radiation

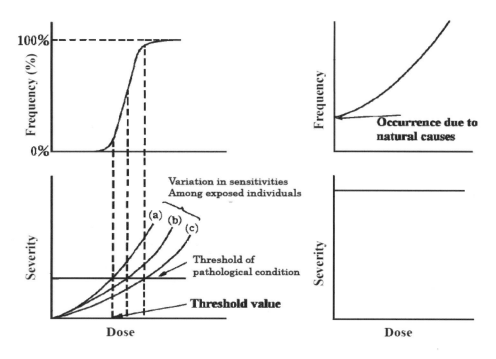

Fig. 43 Schematic expression of variations for occurrence and severity of deterministic effects (left) and stochastic effects (right)

Table 8 Classification of biological effects of radiation

Radiation effects	Somatic effects	Acute effects	Erythema on the skin Loss of hair Reduction of white blood cell Infertility, etc.	Deterministic effects
		Late effects	Cataracts Effects on embryos, etc.	
			Leukemia Cancer	Stochastic effects
	Heritable effects		Metabolism abnormalities Cartilage abnormalities, etc.	

6. Effects from External and Internal Exposures

In radiation protection, external exposure and internal exposure are considered separately. In planning and implementing measures for radiation protection, it is important to distinguish between external exposure and internal exposure - that is, between radiation from radiation sources outside of the body, and radiation from radioisotopes within the body.

When handling sealed radiation sources or operating radiation generators that do not produce radioisotopes, only external exposure needs to be considered. When handling unsealed radioisotopes emitting low-energy beta rays or alpha rays only, internal exposure generally is considered (though this depends also on the method of handling and other conditions). Otherwise, both external and internal exposures should be considered.

In the case of external exposure, strongly penetrating radiation such as gamma rays (or X-rays) or neutrons, is the main concern. In contrast, with internal exposure, radiation with low penetration power, such as alpha and beta rays, are the principal problem.

The activity of a radioisotope taken into the body by inhalation decreases in accordance with its inherent physical half-life T_p. The activity in the body is also reduced by metabolism. Formally, the reduction was assumed to be expressed by a simple exponential function and the "biological half-life" T_b was defined. Thus, the amount of radioisotope in the body is reduced in accordance with the effective half-life T_{eff}, as shown by the following formula:

$$\frac{1}{T_{eff}} = \frac{1}{T_p} + \frac{1}{T_b}$$

At present, more realistic biokinetic models are constructed for most of the radioisotopes and for their physical and chemical forms and, based on these models, conversion coefficients of effective dose per unit intake, "**dose coefficients**", are obtained. Some examples are shown in Table 10.

IV. Biological Effects of Radiation

The intake of radioisotopes by ingestion should theoretically be neglected because eating and smoking are strictly prohibited in any controlled area where unsealed radioisotopes are handled.

Regardless of whether exposure is external or internal, if the received effective dose, or more precisely, committed effective doses for internal exposure, are the same, the degree of effect on the body is assumed to be the same.

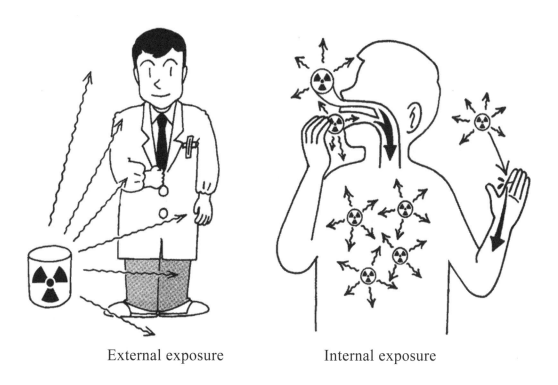

External exposure   Internal exposure

Fig. 44 In implementing measures for radiation protection, it is important to think separately of circumstances under which the body might be subjected to external and to internal exposure.

7. What is Equivalent Dose?

Equivalent dose is the basic dose concept used in radiation protection, the unit for which is the sievert (Sv). With the equivalent dose, regardless of the type of radiation, when the equivalent dose to a particular tissue or organ is the same, the radiation effects to the tissue or organ are assumed to be the same. Equivalent dose can be obtained by multiplying absorbed dose (an actual physical quantity) to tissues and organs by a radiation weighting factor, which is a function of the type and energy of the radiation.

7-1. Index for Assessing Radiation Effect on the Body

Radiation can be easily measured based on physical phenomena, such as ionization and photographic effect. In order to use the measured quantity to assess the effect on the human body, however, it is necessary to use an adjustment value, called a **radiation weighting factor** (w_R), according to the type of radiation. The result obtained by multiplying the absorbed dose by the radiation weighting factor is defined as the **equivalent dose**. Equivalent dose is the basic index for assessing radiation effect on the body.

7-2. Radiation weighting factors as applied to different types of radiation

In its 2007 Recommendations, the ICRP proposed applying the following radiation weighting factors:

        1 for gamma rays, X-rays, electrons and beta rays
        2 for Protons
        20 for alpha rays, and
        2.5~about 20 for neutrons (depending on neutron energy).

Accordingly, equivalent doses corresponding to 1 mGy from gamma rays and 1 mGy from alpha rays are, respectively, 1 mSv and 20 mSv.

7-3. Assessing Equivalent Dose

Equivalent dose, which is an index of the effect on the body, is assessed for both tissue equivalent dose, showing dose at a specific part of the body. Tissue equivalent dose are indirectly measured by monitoring the individuals concerned using various kinds of personal monitors (see III 5).

IV. Biological Effects of Radiation

For purpose of dose assessment for dose limit of equivalent dose in skin, 70 μm dose equivalent should be measured.

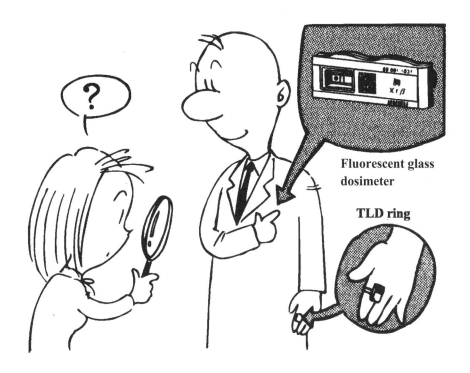

Fig. 45 Measurements of 1cm- and 70 μm depth dose equivalents

8. What is Effective Dose?

Effective dose is a concept quantifying radiation risk. Again, the sievert (Sv) is the unit of effective dose. Equivalent dose for each tissue or organ is multiplied by the appropriate tissue weighting factor w_T (Table 9), and the results for all the tissues and organs are added together to yield the effective dose. The dose limit of 50 mSv/y and 100 mSv/5y for radiation workers, specified in Chapter V, Laws and Ordinances, is expressed in terms of the effective dose.

8-1. Accounting for the radiation sensitivity of each tissue or organ

Sensitivity to radiation varies with the tissue or organ. In other words, even if the equivalent dose is the same, the probability that stochastic effects (incidence of cancer and heritable effects) will appear depends on the tissue or organ that is irradiated. Effective dose was therefore created as a measure by which radiation exposure can be compared, regardless of which parts of the body are exposed, in terms of stochastic effects. The unit is the sievert (Sv), as in the case of equivalent dose. Effective dose is used when planning radiation protection and exposure control.

8-2. Applicable regardless of how exposed

An advantage of the effective dose is that it makes possible to combine and assess together the results of external exposure and internal exposure, or whole-body exposure and **partial-body** exposure.

8-3. Use of dose equivalent quantities for measurements of external dose

Effective dose is the total of all organ equivalent doses multiplied by their respective tissue weighting factors. As it is impossible to measure the effective dose directly, 1-cm dose equivalent is assessed through personal and environmental monitoring.

These two categories of dose equivalent quantities are called **operational quantities** as compared to **protection quantities** for equivalent dose and effective dose. The dose equivalent quantities are defined by the International Commission of Radiation Units and Measurements (ICRU) with simplified source-detector geometry to avoid underestimation of both equivalent dose and effective dose.

IV. Biological Effects of Radiation

Table 9 Tissue weighting factors provided in the 2007 Recommendations of the ICRP

Tissue/Organ	Tissue weighting factor W_T	Tissue/Organ	Tissue weighting factor W_T
Bone marrow (red)	0.12	Liver	0.04
Colon	0.12	Thyroid	0.04
Lung	0.12	Bone surface	0.01
Stomach	0.12	Brain	0.01
Breast	0.12	Salivary glands	0.01
Gonads	0.08	Skin	0.01
Bladder	0.04	Remainder tissues*	0.12
Esophagus	0.04	Total	1.00

*Remainder Tissues(14 in total): Adrenals, Extrathoracic (ET) egion, Gall bladder, Heart, Kidneys, Lymphatic nodes, Muscle, Oral mucosa, Pancreas, Prostate, Small intestine, Spleen, Thymus, Uterus/cervix.

Table 10 Some examples of effective dose coefficients for inhalation

Nuclides	Chemical form	Dose coefficients (mSv/Bq)
^{3}H	Elementary	1.8×10^{-12}
	Methane	1.8×10^{-10}
	Water	1.8×10^{-8}
	Organic compounds except methane	4.1×10^{-8}
	Other compounds	2.8×10^{-8}
^{14}C	Gaseous	5.8×10^{-7}
	Carbon monoxide	8.0×10^{-10}
	Carbon dioxide	6.5×10^{-9}
	Methane	2.9×10^{-9}
^{32}P	Compounds except tin phosphates	1.1×10^{-6}
	Tin phosphates	2.9×10^{-6}
^{60}Co	Compounds except oxides, hydroxides, halogenides, nitrates	7.1×10^{-6}
	Oxides, hydroxides, halogenides, nitrates	1.7×10^{-5}
^{131}I	Elementary	2.0×10^{-5}
	Methyl iodide	1.5×10^{-5}
	Compounds except methyl iodide	1.1×10^{-5}
^{137}Cs	All compounds	6.7×10^{-6}

IV. Biological Effects of Radiation

9. Assessment of Internal Dose

Radioisotopes taken into the body by inhalation or by ingestion are either excreted or remain in the body as radiation sources until the radioisotopes completely decay or are excreted. Dose from such remaining radioisotopes is estimated taking future dose into account. This is called committed equivalent dose to the relevant tissue or organ. The committed effective dose is a total sum of these committed equivalent doses multiplied by the tissue weighting factor for each tissue or organ.

9-1. Exposure from radioisotopes taken into the body

Radioisotopes taken into the body are either excreted from the body mainly in urine and feces, or remain in the body emitting radiation until they completely decay.

9-2. Committed equivalent dose is used as equivalent dose for internal exposure

If exposure continues for five to ten years, for example, it is inconvenient to evaluate and record the dose yearly. Rather, it is more appropriate to think that one receives a dose — including what is estimated to be received in the future

— back at the time when the radioisotope was taken into the body. Because the dose is thus counted in advance, the total integrated dose for the period when the isotopes are in the body is called **committed equivalent dose**, and is used as the measure of internal dose. When the period of integration is not specified for a particular case, it is taken 50 years from the intake for workers and up to 70 years of age for the public.

9-3. Assessments of committed equivalent dose and effective dose

To calculate committed equivalent dose when radioisotopes are taken into the body, a knowledge of how radioisotopes behave in the body is required. There are a lot of studies in this field resulting in the development of biokinetic models including those of the respiratory system, gastrointestinal system, et al. They

were used by the ICRP to publish the numerical data of effective dose for unit activity intake (Sv/Bq), called **dose coefficients** for many radioactive nuclides and their physical and chemical forms. When the amount of activity of a radioactive nuclide taken by a person becomes known, the committed effective dose for the nuclide can be calculated by multiplying the activity with the relevant dose
coefficient.

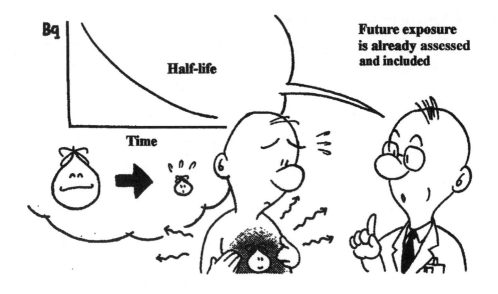

Fig. 46　Committed equivalent dose

IV. Biological Effects of Radiation

10. Natural Radiation and Artificial Radiation

Between natural and artificial radiations, there is not any difference in biological effects on the body.

From the point of view of the effects on human body, whether it is natural radiation or artificial radiation makes no difference. Natural and artificial radiations of the same type and with the same energy have the same effect on the body.

Fig. 47 Natural and artificial radiations

People living on the earth are inevitably exposed to natural and artificial radiation. Typical values of dose to them from various sources are shown in Fig. 48 and Table 11.

The effective dose to Japanese people as a result of the Chernobyl accident was 0.005 mSv for the year following the accident. Total lifetime effective dose is estimated at 0.006 mSv or less.

('Chernobyl Radioactivity and Japan,' 1989, and 'Proceedings of the 8th International Congress of Radiation Research,' 1987)

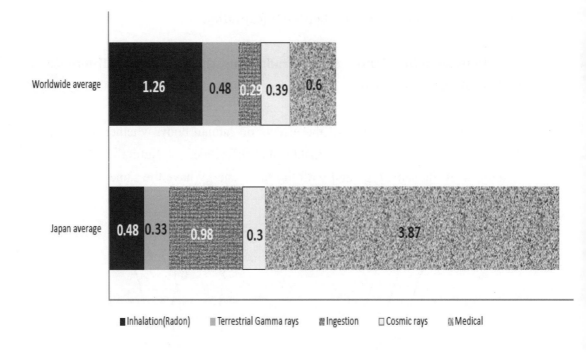

Fig. 48.   Average of annual dose (mSv) in worldwide and Japan
('Environmental Radiation in Life' by the Nuclear Safety Research Association 2011)

Table 11   Typical estimated values of effective dose to people in the world (UNSCEAR 2008)

Average per person (mSv)

Natural background 2.4 mSv	Inhalation (Radon)	1.26
	Terrestrial gamma rays	0.48
	Radionuclides in tissue	0.29
	Cosmic rays	0.39
Artificial exposure 0.6 mSv	Medical exposure (diagnosis)	0.6
	Atmospheric weapon testing	0.005
	Occupational exposure	0.005
	Chernobyl accident	0.002
	Nuclear power generation	0.0002

IV. Biological Effects of Radiation

11. Radiation and Health Effects

Effects that can clearly be attributed to radiation do not occur unless one is exposed to a large amount of radiation due to, for example, an accident.

In addition, a person's recovery from the effects of radiation exposure depends on the dose received. When the dose is small, effects are temporary and disappear after some time. The higher the dose becomes, however, the more serious the effects are, and the more difficult to recover from them.

For example, assume that the skin is exposed partially in an accident to a large amount of weakly penetrating radiation such as beta rays or soft X-rays. If the dose is 3 to 5 Gy*, the skin will become red and hair will fall out. These effects are temporary, however, and the skin will return to normal in several months. If the dose is 5 to 10 Gy, blisters will form, as when one is burned, but here, too, the skin will recover in several months. On the other hand, if the dose exceeds 25 Gy, the exposed skin will become ulcerated and will never return to its original condition. The surface of the skin (the outer layer) will become thinner, and the tissues beneath will become harder and thicker, and the appearance will become blotchy.

The important thing to remember is that the dose needed to cause such deterministic effects is far higher than the dose limit for the skin (0.5 Sv/y) for radiation workers.

Additionally, there is no evidence, that radiation workers catch colds more easily or become physically weaker than before they were exposed to radiation during normal operations. Easily catching colds and being weak physically are related to the body's immune mechanism. Immunity involves various kinds of cells, such as those of the lymphatic system and bone marrow, and changes in these cells do not occur unless one is exposed to radiation in amounts large enough to cause the white blood cell count to decrease. Thus, there is no risk of affecting the immune system when engaging in normal radiation work.

---

* Here absorbed dose (Gy) is used instead of equivalent dose (Sv), because tissue weighting factor recommended by the ICRP is applicable only to low dose where stochastic effects are relevant.

Fig. 49 Radiation and health effects

IV. Biological Effects of Radiation

12. Exposure Categories

12-1. Occupational exposure

There are many jobs involving radiation, and the number of people involved varies depending on the task and the type of radiation or radioactive material being handled. The average exposure dose per person for various occupational fields (excl. Nuclear Power Plant) was surveyed by the Council on Personal Dosimetry Service in 2014, and the results are shown in Table 12. As seen in the Table, Non-destructive Testing workers received the largest exposure dose, with an average dose of 0.45 mSv. In contrast, workers in research and education area received the lowest dose at 0.027 mSv. An investigation by the Radiation Effects Association reported that the annual average effective dose per person among nuclear power plant and related contract workers was 1.5 mSv. All these kinds of exposure are called **occupational exposure**.

Table 12 Annual average dose per person (mSv) (Japan FY 2014)

Occupational Field	Effective Dose (mSv)	Number of workers
General Medicine	0.41	333,930
Dental Medicine	0.028	22,078
Veterinary Medicine	0.038	14,108
Non-destructive Testing	0.45	3,736
General Industry	0.064	69,162
Research and Education	0.027	68,856
Nuclear Power Plant	1.5	74,644

12-2. Medical Exposure

Exposure of patients during medical therapy and examinations varies greatly between adults and children, among countries and regions, and by individual. The magnitude of these variations can easily be grasped simply by thinking of the difference between receiving, or not receiving, radiation cancer therapy. There are increasing opportunities to be exposed to radiation for medical diagnoses - dental X-rays, stomach and chest examinations, computerized X-ray tomography (CT scans), etc. In addition, radiation is used to cure cancer. These kinds of exposure are called **medical exposure**. Medical exposure also includes the exposure of volunteers for biomedical research.

12-3. Public Exposure

All other exposures are taken to be **public exposure**. Ambient exposure includes exposure by cosmic rays, radiation from natural radioactive elements in the earth and in the atmosphere, fallout from past nuclear testings, and radioactive materials discharged from nuclear power plants currently in service. The level of natural radiation, in particular, varies greatly depending on area and time, which results in substantial differences in exposure dose.

IV. Biological Effects of Radiation

13. Risk

The use of radiation and nuclear power brings various benefits to our daily lives. Nuclear power generation, for example, accounts for about 30 percent of total electricity generation in Japan (∼February 2011) and is essential to the stability of the nation's power supply. The use of radiation in medicine contributes significantly to keeping us healthy. Radiation and nuclear power thus contribute positively to human society; yet exposure to radiation itself is considered harmful to the body. Harmful effects, however, do not necessarily appear whenever one is exposed to radiation. In particular, the dose a person receives daily is so small that it is not clear whether there are any harmful effects at all.

The degree of harmful effects that may occur as a result of usage of radiation specifically, the probability of the effects occurring, is called the risk. Risk does not mean that harmful effects will occur; it just means the possibility that they will occur.

Fig. 50   Double-edged sword

IV. Biological Effects of Radiation

14. Limits on Personal Exposure

For radiation workers exposure is controlled so as to remain below a certain level. This level is the dose limit. The tissue equivalent dose limit for deterministic effects and the effective dose limit for stochastic effects are specified in the Act and Ordinances. Radiation dose is controlled so as not to exceed either of these values.

14-1. Dose limit for radiation workers

The dose for radiation workers should be controlled so as not to exceed the specified limits for either effective dose or tissue equivalent dose. The effective dose limit and tissue equivalent dose limit are both controlled based on one year periods starting April 1, except in the case of women. Values are shown in Table 13.

Table 13 Effective dose limits and tissue equivalent dose limits for radiation workers

Item		Radiation Worker
Effective dose limits		(1) 100 mSv/5 years[*1] (2) 50 mSv/year[*2] (3) Women[*3] 5 mSv/3 months[*4] (4) Pregnant women 　　The internal exposure limit set for this woman is 1mSv for a period of time starting from the time when her employer and others are informed of her pregnancy by her reporting or any other means up to the time of the delivery of the baby
Equivalent dose limits	(1) The lens of the eye	150 mSv/year[*2]
	(2) Skin	500 mSv/year[*2]
	(3) Abdomen surface of pregnant women	2 mSv for a period of time starting from the time when her employer and others are informed of her pregnancy by her reporting or any other means up to the time of the delivery of the baby

* 1: Each period of time derived from the division of years from April 1, 2001 into increments of 5 years

* 2: The time span of 1 year with April 1 as the starting date

* 3: Excluding those who were diagnosed as being unable to achieve a pregnancy, those who made an offer of having no intention of becoming pregnant in writing, and those who are currently pregnant

* 4: Three month periods each with the starting date set on April 1, July 1, October 1 and January 1

IV. Biological Effects of Radiation

14-2. Exposure not counted in dose limit

According to the Act on Prevention of Radiation Hazards, electrons and X-rays with energy less than 1 MeV are not subject to regulation, but exposure dose from such radiation must be included in total personal exposure. Exposures of patient due to X-ray diagnosis, diagnosis with nuclear medicines, radiation therapy, and exposures of radiation public due to natural radiation are not subject to regulation with respect to the dose limits.

14-3. Equivalent dose limits other than those mentioned above

Dose limit for radiation workers engaging in emergency operations in radiation facility is stipulated as 100 mSv in the Law; this is limited to men and to women who are diagnosed as infertile.

14-4. Total of external exposure and internal exposure

When both external and internal exposures occur, the effective dose of each is separately evaluated in terms of the fraction of the effective dose limit. The combined limit for the two types of exposure is any combination of the two fractions that total 1.

V. The Act and its Ordinances

1. Act on Prevention of Radiation Hazards due to Radioisotopes, etc.

"Act on Prevention of Radiation Hazards due to Radioisotopes, etc. including its Ordinance and notifications etc. (the Act and relevant Ordinances)" apply uniformly regardless of the nature of the establishment. It is appropriate to think of them as stipulating the minimum standards that the establishments and the radiation workers must observe, and, accordingly, each establishment, while of course meeting the legally prescribed standards, should implement radiation control appropriate for its own activities.

1-1. Why is control based on the Act and relevant Ordinances necessary?

Radiation and radioactive materials are causes of hazardous effects. When people attempt to use them, their use should always be controlled.

1-2. Historical background for the Act and relevant Ordinances concerning radiation hazards in Japan

Radiation, especially X-rays, came into use relatively early in the medical and research fields. It wasn't until the mid-1950s that the use of radiation and radioisotopes in areas other than medicine and research began. In response to the new developments, the Atomic Energy Basic Act and the Act on Prevention of Radiation Hazards due to Radioisotopes, etc. and relevant Ordinances were passed in 1955 and 1957, respectively. Both have been revised many times since then, based on the latest scientific and technological knowledge. Because the Act on Prevention of Radiation Hazards due to Radioisotopes, etc., was drafted and modified in line with recommendations of the ICRP, its provisions on radiation control are not significantly different from those in other countries.

1-3. Characteristics of the Act

The purpose of the Act (on Prevention of Radiation Hazards due to

V. The Act and relevant Ordinances

public and should be obeyed when radiation or radioisotopes are used. In order to attain this purpose, the Act provides for the regulation of facilities (requirements for facilities) and for the regulation of people's actions, etc. (requirements for actions).

Fig. 51 Role of the Act and relevant Ordinances

94

2. Hierarchy of the Act and relevant Ordinances

Laws are enacted by the Diet, which is the national legislative body. Various administrative bodies are then empowered to establish statutory instruments – government Order (Cabinet Order for Enforcement of the Act), ministerial ordinances (NRA Ordinances for Enforcement of the Act), and public notices issued by any minister directly to the public. In this chapter these are collectively called Laws and Ordinances.

2-1. Relationship among the Act and relevant Ordinances

An example of the relationships among the Act and relevant Ordinances is shown in Table 14. The Act is enacted by the Diet, establishing fundamental principles. Details are then provided in the Cabinet Order and the NRA Ordinance for enforcement of the Act, and notifications.

2-2. Administrative Guidance

Despite the existence of the Act and relevant Ordinances, many actual situations involving nuclear power and radiation may not be adequately dealt with by specific provisions. Thus, administrative bodies sometimes provide guidance in the form of guidelines and notifications, in order to attain the administrative purpose.

2-3. Rules in each establishment

Each establishment is required to establish **Radiation Hazards Prevention Program**, in accordance to the Act and relevant Ordinances. Under the Act on the Regulation of Nuclear Source Materials, Nuclear Fuel Materials and Reactors, it is called **Operational Safety Program**.

Many establishments, moreover, in addition to their Programs, maintain their own radiation control manuals, which can be revised as appropriate at any time.

V. The Act and relevant Ordinances

Table 14 Hierarchy of the Act and relevant Ordinances

The Act and relevant Ordinances	Diet	Act	Act on Prevention of Radiation Hazards due to Radioisotopes, etc. Radioisotopes, etc.(duplicate, deleted)
	Cabinet	Cabinet Order	Cabinet Order for Enforcement of the Act Prevention of Radiation Hazards due to Radioisotopes, etc.
	NRA*	Ordinance	NRA Ordinance for Enforcement of the Act Prevention of Radiation Hazards due to Radioisotopes, etc.
	NRA*	Notifications	Notification to Specify Standards for the Amount, etc. of Radioisotopes
NRA*		Announcement	Safety for Incineration of Waste Liquid Scintillator
Authorized Operators		Internal practices	Radiation Hazards Prevention Program

*NRA : Nuclear Regulation Authority

3. **Laws and Ordinances Concerning Radiation**

In order to prevent radiation hazards for radiation workers and the general public, while furthering the use of radiation and radioisotopes, many Laws and Ordinances have been enacted from various points of view, and the facilities where radiation and radioisotopes are used are regulated by these Laws and Ordinances.

3-1. **Laws and Ordinances concerning radiation**

Laws and Ordinances concerning radiation include the **Atomic Energy Basic Act**, which deals with the development and use of nuclear power; the **Labour Standards Act**, concerned with the protection of radiation workers at radiation facilities; the **Medical Care Act**, regulating the use of radiation for diagnosis and therapy; and the **Act for Ensuring the Quality, Efficacy, and Safety of Drugs and Medical Devices**, on the production of radiopharmaceuticals.

3-2. **Laws and Ordinances to promote the use of atomic energy**

The principal Law governing promotion of research and development, and the use of nuclear power, is the Atomic Energy Basic Act. Under the Law, research and development, and use of atomic energy, are limited to peaceful purposes, with assured safety, independent operations, government oversight results made available to the public, and a commitment to international cooperation.

The Atomic Energy Basic Act also provides for the separate enactment of the Act on Prevention of Radiation Hazards due to Radioisotopes, etc. to prevent such hazards and ensure public safety; and of the Act on the Regulation of Nuclear Source Material, Nuclear Fuel Material and Reactors (shortly, **the Nuclear Reactor and Fuel Regulation Law**), to regulate safety control in regard to nuclear materials and reactor operations.

V. The Act and relevant Ordinances

3-3. Laws and Ordinances to protect workers

The Labour Standards Act was enacted to protect workers. Under that Law, the Industrial Safety and Health Act was enacted to provide for the safety and health of workers at workplace. In the enforcement of that Law, radiation-related matters are stipulated in **Ordinance on Prevention of Ionizing Radiation Hazards**.

4. What is the Act on Prevention of Radiation Hazards due to Radioisotopes, etc. (the Act)?

Pursuant to the Atomic Energy Basic Act, the Act was enacted to prevent radiation hazards and ensure public safety, by regulating the use of radiation and radioisotopes, and management of materials contaminated by radioisotopes. Originally enacted in 1957, it has been revised in accordance with recommendations of the ICRP.

4-1. Requirements for Facilities

Prior to the use of radiation or radioisotopes, permission by or notification to the Nuclear Regulation Authority (NRA) is required. Upon granting permission, the NRA confirms that the location, structure, facilities to be used, and storage and waste management facilities meet the technical standards (standards for facilities) prescribed in the Act and relevant Ordinances. Even after permission is granted, the NRA conducts regular periodic inspections of all facilities handling radioisotopes in excess of a defined quantity, to see that the technical standards are maintained (regular periodic inspections, etc.).

Table 15 Standards for facilities requiring permission or notification

Permitted User	Registered User
The use of radioisotopes or radiation generators (excluding those for which the notification requirements apply)	The use of radiation sources of the quantity exceeding the exemption limit but not exceeding 1,000 times the lowest limit (quantity)

4-2. Requirements for Actions

Those who are given permission to use, or who submit a notification on the use of radioisotopes or radiation, must comply with a series of standards for safety control, in order to prevent radiation hazards. Specific standards apply in the following areas: education and training of workers who enter controlled areas; measurements of personal dose and medical examinations for such workers; measurements of dose rate at places where there is risk of radiation hazard; and use, repacking, storage and transportation of radioisotopes and waste management.

4-3. Local Rules on Radiation Safety Control

Detailed rules on radiation safety control, in accordance with the conditions and circumstances of the establishment, must be provided in each establishment's Radiation Hazards Prevention Program.

5. Laws and Ordinances to Protect Workers

Ordinance on Prevention of Ionizing Radiation Hazards was established under the Industrial Safety and Health Act, solely to protect workers at working sites from radiation hazards. Similar regulations are established under different Laws for seamen, government employees, self-defense-force members, miners, and others.

5-1. The Ordinance of the Ministry of Health, Labour and Welfare: "The Ordinance on Prevention of Ionizing Radiation Hazards"

The Ordinance on Prevention of Ionizing Radiation Hazards was established under the Industrial Safety and Health Act deleted, providing for regulations based on the principle that employers (operators) must try to minimize ionizing radiation to which workers are exposed in the course of their jobs.

An industrial physician and a safety supervisor must be appointed under the regulations. Safety supervisors deal with technical matters associated with safe operations of X-ray apparatuses such as X-ray analyzers and X-ray diffraction devices.

These regulations, unlike the Radiation Hazards Prevention Act, cover X-rays and electrons with energy less than 1 MeV, reactor operations and the mining of nuclear source materials.

5-2. Working Environment Measurement Act

This Law, together with the **Industrial Safety and Health Act**, stipulates qualifications of experts and organizations for monitoring.

5-3. Rules of the National Personnel Authority

The Rules of the National Personnel Authority 10-5 (Prevention of Radiation Hazards for Employees) were established under the **National Public Service Act**, to protect regular government employees from radiation hazards. The contents are almost the same as those of the Ordinance on Prevention of Ionizing Radiation Hazards that are not applicable to government officials.

V. The Act and relevant Ordinances

Fig. 52   Working Environment Measurement Act and Ordinance on Prevention of Ionizing Radiation Hazards

6. Who are Radiation Workers?

Those who deal with radiation or radioisotopes in controlled areas are legally called "radiation workers". Because radiation workers handle radiation and radioisotopes, which are considered to be harmful materials, such workers must fully understand the basic concepts regarding radiation and its deleterious effects, and observe the provisions in the Radiation Hazards Prevention Act and its Ordinances and in the Radiation Hazard Prevention Program at their facilities.

6-1. Radiation workers

Those who handle radiation generators or radioisotopes mainly in controlled areas are called "**radiation workers**" under the Act on Prevention of Radiation Hazards due to Radioisotopes, etc..

6-2. Why is it necessary to specifically identify "radiation workers"?

Radiation and radioisotopes are considered harmful, and should not be handled by just anybody. Only those who have completed specific steps to work in controlled areas are allowed to use radiation or radioisotopes − i.e., to become as radiation workers. Radiation workers, in the course of their handling of radiation and radioisotopes, are concurrently subject to necessary regulations control (education and training, exposure control and health control).

6-3. Steps in becoming a radiation worker

Only those who 1) have had required education and training and 2) have had medical examinations, are allowed to enter controlled areas as radiation workers. Education and training are given to radiation workers to ensure that they have the enough knowledge and skills required of radiation workers. Medical examinations are required to determine if the individual is medically suitable for assignments involving the handling of radiation and radioisotopes.

If a radiation worker has experience handling radiation at, for example, another establishment, his or her previous dose records must be checked.

V.　The Act and relevant Ordinances

6-4.　Regulations

Workers designated as "radiation workers" may handle radiation and radioisotopes in controlled areas according to all applicable rules and procedures.

Radiation workers receive on a regular basis:　1) education and training, 2) medical examinations, and 3) personal monitoring in controlled areas. Radiation workers must be informed of the results of their personal monitoring each time, and they themselves should make a habit of confirming such results.

7. Who are Radiation Protection Supervisors?

In order to conduct appropriate and thorough radiation safety control at an establishment, it is necessary to have a management structure or mechanism in place that is responsible for radiation protection control within the establishment. In such a situation, the radiation protection supervisor is playing a key role. The radiation protection supervisor makes the radiation protection program for the establishment, taking its nature and specific circumstances into consideration, and oversees operations to make sure the program are implemented properly.

7-1. Who are radiation protection supervisors?

Radiation protection supervisors are the people responsible for supervising efforts to prevent radiation hazards. They are appointed at any establishment where radiation or radioisotopes are handled in the course of using, selling, leasing, or Ieasing of them and waste management.

7-2. Role of radiation protection supervisors in the establishment

Radiation protection supervisors are responsible for supervising all matters associated with radiation control and radiation safety in their establishment. Radiation workers must follow the instructions of the radiation protection supervisor concerning radiation safety control. The employer, too, must pay serious attention to the decisions of the radiation protection supervisor on matters necessary for radiation safety control.

7-3. Categories of radiation protection supervisors

To be qualified as a radiation protection supervisor, a person must complete required training courses and pass a national examination. There are three classes of radiation protection supervisors — first class to third class. Which class is required by a particular establishment depends on the kind and level of radiation or radioisotopes handled. Radiation protection supervisors must attend seminars, etc., on an on-going basis, in order to always have the latest information in the field of radiation protection.

V. The Act and relevant Ordinances

7-4. Cooperation with radiation protection supervisor

The radiation protection supervisor is responsible for supervising radiation safety control at an establishment. Good radiation safety control will be implemented by good communication among radiation protection staff and radiation workers under a leadership of the radiation protection supervisor (see Ⅲ1).

Fig. 53　Radiation protection supervisor

8. What is Radiation under Laws and Ordinances?

"Radiation" as used in Laws and Ordinances does not mean radiation as the word is understood in physics; rather, it means radiation as specified in the Atomic Energy Basic Act. Among the many kinds of electromagnetic waves and particles, the Act defines radiation as alpha rays, gamma rays and other radiations with the ability to directly or indirectly ionize the air.

8-1. There are various types of radiation

Radiation broadly means all electromagnetic waves and particles. X-rays from X-ray generators, alpha and beta rays emitted during the decay of atomic nuclei, and gamma rays, are all well-known types of radiation. Particles emitted during nuclear reactions, mutual transformations of elementary particles, and cosmic rays are also types of radiation.

Legally, however, radiations generated artificially and that have the ability directly or indirectly to ionize air are defined as "radiation", as shown below.

The radiation defined in the Cabinet Order for the Defrnition of Nuclear Source Material, Nuclear Fuel Material and Nuclear Reactor is:
1) Alpha rays, deuterons, protons and other heavy charged-particles, and beta rays;
2) Neutrons;
3) Gamma rays and characteristic X-rays (limited to characteristic X-rays generated at the time of electron capture); and
4) Electrons and X-rays with energy of 1 MeV or more.

8-2. Definitions of radiation in other Laws and Ordinances

According to the NRA Ordinance for Enforcement of Radiation Hazards Prevention Act, electrons and X-rays with energy of 1 MeV or greater are classified as "radiation", whereas the Medical Care Act regulates X-ray generators with a tube voltage of 10 kV or more. Also, under the Ordinance on Prevention of Ionizing Radiation Hazards of the Ministry of Health, Labour and Welfare (MHLW), although the Ordinance does not specify an energy limit, the Minister of Health, Labour and Welfare has decided that X-ray generators

V. The Act and relevant Ordinances

with tube voltage of 10 kV or more must be regulated for the purpose of radiation protection.

8-3. Radiation subject to dose evaluation

Electrons and X-rays less than 1 MeV are not "radiation" as defined by the Radiation Hazards Prevention Act and the Nuclear Reactor Regulation Act, whereas they are included in the Labuor Standards Act.

However, when dose is assessed, exposure to such radiation is to be added to exposure from other legally defined "radiation".

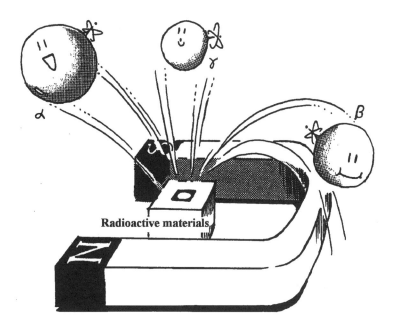

Fig. 54　Various types of radiations

9. What are Radioisotopes?

Legally, "radioisotopes" are isotopes emitting radiation, compounds of them, and materials containing them (including those equipped in instruments), in quantities or concentrations exceeding those stipulated by the NRA.

Such quantities or concentrations are called **"exemption limit"**, because radioisotopes below the limit quantity are exempted from regulations for radioisotopes.

Table 16 shows examples of the exemption limit for radioisotopes frequently used in laboratories and factories. This means that anyone who has no experience to deal with radioisotopes can have or use 10^8 Bq or 10^5 Bq/g of tritium of any form without permission by or notification to the NRA.

It should be noted that the exemption level is not the activity level for disposing of radioactive wastes without permission nor notification (**clearance**). Levels for clearance of radioactive wastes are not yet stipulated by the NRA.

Table 16 Quantities and Concentrations of Main Radioisotopes above which are Subject to Regulations

Radioisotopes	Activity (Bq)	Activity concentration (Bq/g)
^{3}H	1×10^9	1×10^6
^{14}C	1×10^7	1×10^4
^{32}P	1×10^5	1×10^3
^{60}Co	1×10^5	1×10^1
^{131}I	1×10^6	1×10^2
^{137}Cs	1×10^4	1×10^1

V. The Act and relevant Ordinances

10. What are Radiation Generators?

At present, many kinds of radiation generators are in use. There are substantial differences in capacity among such equipment, and the applicable Laws differ. Under the Act on Prevention of Radiation Hazards due to Radioisotopes, etc., several types of devices, including cyclotrons and linear accelerators, are subject to regulation.

10-1. Radiation generators under Laws and Ordinances

"Radiation generators" covered by the Radiation Hazards Prevention Act and its Ordinances includes such devices as cyclotrons and synchrotrons generating radiation by accelerating charged particles, as shown in the Table 17. Ordinary X-ray generators for medical examinations are not included because the X-ray energy is below 1 MeV. Also, when the maximum equivalent dose rate at a point 10 cm away from the surface of the equipment is 600 nSv/h or less, such equipment is not included.

10-2. Use of radiation generators

At the entrance to a room where radiation generators are operated, there must be an automatic indicator which shows whether the generator is in operation or not, and additionally there must be interlock systems at the entrance.

Table 17 Radiation generators under the Radiation Hazards Prevention Act*

Radiation generators
1. Cyclotron
2. Synchrotron
3. Synchro-cyclotron
4. Linear accelerator
5. Betatron
6. Van de Graaff accelerator
7. Cockcroft-Walton accelerator
8. Other types of radiation generators by accelerating charged particles, specified by the NRA: transformer-type accelerator, microtron, and plasma generating equipment capable of attaining a critical plasma condition for the D-T reaction

*Equipment for which the 1 cm depth dose equivalent at a point 10cm away from the surface of the equipment is 600 nSv/h or less are excluded.

11. What are Sealed Radioisotopes?

Radioisotopes are classified into two categories for the purpose of control – sealed radioisotopes (sealed sources) and unsealed radioisotopes (unsealed sources). Included in the sealed sources are sources for radiation therapy, industrial sources, calibration sources, and electron-capture detectors for gas chromatography. Sealed sources present no danger in regard to contamination, but external exposure must be a concern.

11-1. Safe handling of sealed sources

Sealed radioisotopes are basically used in circumstances that involve no danger of their being opened or damaged under normal conditions, and they present no risk of the contents being scattered and causing contamination as a result of leakage, permeation, etc. It is, however, always important to prevent overexposure by taking appropriate measures for protection, regardless of the size or nature of the source. Also, leak testing should be done regularly, though it is not required legally.

One of uses of sealed sources is in "radioisotopes-equipped devices with design certificate." These are instruments that the NRA has recognized as meeting technical standards in their mechanisms for the prevention of radiation hazards. One example is an electron capture detector (ECD) for gas chromatography. In handling an ECD, there is no danger of overexposure and the controlled area does not extend outside the cabinet of the chromatography device, but it is important to always control the whereabouts of the sources.

11-2. Management of disused sealed sources

Management of sealed sources that are no longer needed may be different from the handling of d i s u s e d unsealed radioisotopes waste because a radioisotope contained in a sealed source is usually high in activity and has a long half-life. Therefore, recovery of disused sealed sources, including those installed in devices, should be asked of the manufacturer or of the Japan Radioisotope Association.

V. The Act and relevant Ordinances

Fig. 55 Return of sealed sources

12. What is the Controlled Area?

At any facility where radiation or radioisotopes are dealt with, any area where there is a possibility that the level of radiation or the concentration of radioisotopes in air or on surface might rise above a specified level is designated a controlled area, and access to that area must be restricted. When unsealed radioisotopes are used, they must be used in a "working room", where measures for cleaning the exhaust air, the floor and the wall are provided.

12-1. Radiation levels in controlled areas

At frequently entered places (e.g., "working rooms") within controlled areas, in order to maintain total effective dose from external and internal exposure at under 1 mSv per week, shielding material should be installed and working hours restricted when necessary.

In addition, outside the boundary of the controlled area, effective dose from external radiation must not to exceed 1.3 mSv per 3 months. Radiation levels in areas where people live are requested to be the same as those at the boundary of the site of establishment, i.e., 250 µSv per 3 months.

The radiation levels to be maintained are shown in Table 18. At the entrance of a controlled area, a notice on important rules and cautions in the area must be posted.

12-2. Radiation work in controlled areas

When unsealed radioisotopes are handled at a facility, it must be done in a specifically provided "working room". Because such a "working room" is a frequently entered place within a controlled area, the radiation level is controlled not to exceed 1 mSv per week.

12-3. Contamination inspection room

At any facility where unsealed radioisotopes are used, a contamination inspection room has to be provided near the entrance/exit regularly used by workers. In the contamination inspection room, there are washing facilities, a place to change clothes, radiation measuring equipment, etc. When workers leave the controlled area where radioactive contamination may occur (the working room), they must confirm that their bodies are not contaminated.

V. The Act and relevant Ordinances

When skin contamination is detected, it must be removed (decontaminated) using the washing facilities. If any significant contamination still remains on the skin after the decontamination procedure as instructed, a radiation protection staff member shall be notified of the matter.

Table 18 Maximum Permissible Radiation Level to be maintained in and around a controlled area

Area/Item	Frequently entered place within a controlled area	Outside boundary of a controlled area
Effective dose from external radiation	1 mSv per week	1.3 mSv per 3 months
Concentration of radioisotopes in the air	Average concentration for 1 week: as mentioned in column 1 of Table 1 attached to the Notice	Average concentration per 3 months: 1/10 or less that mentioned in column 5 of Table 1 of the Notice
Concentration of surface contamination	Radioisotopes emitting alpha rays: 4 Bq・cm^{-2} or less; Radioisotopes not emitting alpha rays: 40 Bq・cm^{-2} or less	1/10 of the values in the left

13. What is meant by "Radiation Levels outside Controlled Areas"?

In order to prevent hazards to the general public and non-radiation workers from radiation and radioisotopes, limits for radiation levels apply to various areas other than controlled areas. Pursuant to the Radiation Hazards Prevention Act and its Ordinances, limits for radiation levels around the controlled areas and at living quarters within the site are set.

Radiation levels are also regulated at the boundary of the establishment, to ensure the safety of the general public, separately from regulations for the safety of radiation workers working in controlled areas. Limits for radiation levels outside the controlled areas of an establishment are shown in Table 19.

Table 19   Limits for radiation levels at various locations outside controlled areas

Item	Locations
Effective dose from external and internal exposure	Outside the boundary of an establishment: 250 μSv for 3 months; Living quarters within the site: 250 μSv for 3 months; Patients rooms in hospitals and clinics: 1.3 mSv for 3 months
Concentration of radioisotopes in the exhaust air or in the atmosphere	Average concentrations for 3 months are the same as those mentioned in column 5 of Table 1 attached to the Notice
Concentration for liquid waste and drainage	Average concentrations for 3 months are the same as those mentioned in column 6 of Table 1 attached to the Notice

V. The Act and relevant Ordinances

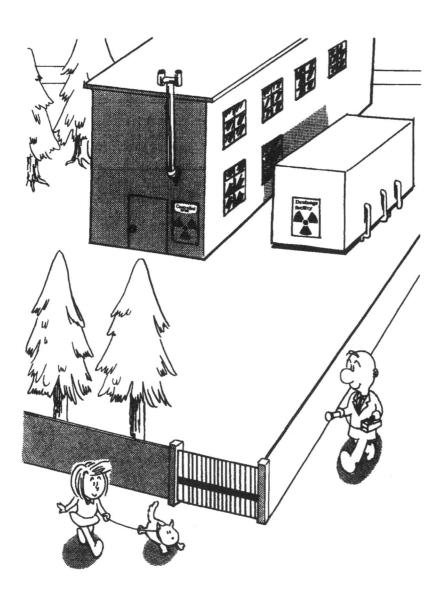

Fig. 56 Boundary of an establishment

VI. Radiation Hazards Prevention Program

1. What is the Radiation Hazards Prevention Program?

The Radiation Hazards Prevention Program is the basis for radiation protection at a establishment. In the Program, the basic concepts of radiation protection at the establishment, the safety control structure, and specific practices that radiation workers must adhere to, are all spelled out. The operating management of the establishment, the radiation protection supervisor, radiation protection staffs and radiation workers must thoroughly understand and observe the Radiation Hazards Prevention Program and observe them. For radiation workers, an explanation of the Radiation Hazards Prevention Program is required to be given prior to the actual start of work in the course of education and training.

1-1. The Radiation Hazards Prevention Program

The Radiation Hazards Prevention Program is the specific program of rules specific rules for the implementation of radiation safety control and the prevention of radiation hazards at each establishment.

The Radiation Hazards Prevention Program is the pledge regarding radiation control at the establishment. The responsibilities of the operating management of the establishment, the radiation protection supervisor, radiation protection staff members, and radiation workers are prescribed.

Items legally required to be included in the Radiation Hazards Prevention Program are listed below.

1-2. Relationship between radiation workers and radiation protection rules

Because the rules at each establishment are based on the Act on Prevention of Radiation Hazards due to Radioisotopes, etc., radiation workers are not required to know all the details of the Act itself. But they must fully understand all items in the Radiation Hazards Prevention Program relating to what they themselves must do. For this reason, lectures on the rules are included within the education and training curriculum provided prior to the worker's handling of radiation or radioisotopes for the first time.

VI. Radiation Hazards Prevention Program

Items to be covered in the Radiation Hazards Prevention Program

(1) Jobs and organizations relating to radiation workers;

(2) Jobs and organizations relating to the radiation protection supervisor and other personnel who are engaged in safety control;

(3) Nomination of acting radiation protection supervisor;

(4) Maintenance and management of radiation facilities;

(5) Inspection of radiation facilities (controlled areas);

(6) Uses of radioisotopes and radiation generator;

(7) Use, repacking, storage and transportation of radioisotopes and waste management;

(8) Measuring, recording and record keeping of radiation dose, etc.;

(9) Education and training required for radiation protection;

(10) Medical examinations;

(11) Health measures required for those who suffer or may suffer from radiation hazards;

(12) Record entries and custody of records;

(13) Safety procedures for disasters such as earthquake and fire;

(14) Safety procedures for emergencies;

(15) Reporting of conditions about radiation safety control; and

(16) Other items necessary for prevention of radiation hazard

Basic Knowledge of Radiation and Radioisotopes

© Japan Radioisotope Association, 2016

First Edition 1997
Revised Edition 2003
Third Edition 2005
Fourth Edition 2016

Published by Japan Radioisotope Association

URL http://www.jrias.or.jp
E-mail syuppan@jrias.or.jp
Address 28-45, Honkomagome 2-chome,
Bunkyo-ku, Tokyo 113-8941, JAPAN

Basic Knowledge of Radiation and Radioisotopes
(Scientific Basis, Safe Handling of Radioisotopes and Radiation Protection)

1997年11月21日	初版発行
2003年 9 月 8 日	改訂版発行
2005年 6 月30日	第 3 版発行 (2005年6月に施行された放射線障害防止関係法令に準拠)
2008年 2 月29日	第 3 版 2 刷発行
2016年 2 月 1 日	第 4 版発行

編　集　　公益社団法人　日本アイソトープ協会
発　行

〒113-8941　東京都文京区本駒込二丁目28番45号
　　TEL　　+81-3-5395-8082
　　FAX　　+81-3-5395-8053
　　E-mail　syuppan@jrias.or.jp
　　URL　　http://www.jrias.or.jp

発 売 所　丸善出版株式会社

© Japan Radioisotope Association, 2016　Printed in Japan

印刷・製本　株式会社 アイワエンタープライズ

ISBN978-4-89073-253-1　C2040